Goddesses, Elixirs, and Witches

Goddesses, Elixirs, and Witches

Plants and Sexuality throughout Human History

John M. Riddle

palgrave
macmillan

GODDESSES, ELIXIRS, AND WITCHES

First published in 2010 by
PALGRAVE MACMILLAN®
in the United States—a division of St. Martin's Press LLC,
175 Fifth Avenue, New York, NY 10010.

Where this book is distributed in the UK, Europe and the rest of the world, this is by Palgrave Macmillan, a division of Macmillan Publishers Limited, registered in England, company number 785998, of Houndmills, Basingstoke, Hampshire RG21 6XS.

Palgrave Macmillan is the global academic imprint of the above companies and has companies and representatives throughout the world.

ISBN: 978–0–230–61064–4

Library of Congress Cataloging-in-Publication Data is available from the Library of Congress.

A catalogue record of the book is available from the British Library.

Design by Newgen Imaging Systems (P) Ltd., Chennai, India.

First edition: February 2010

10 9 8 7 6 5 4 3 2 1

Printed in the United States of America.

Contents

Acknowledgments vii

List of Abbreviations ix

Introduction 1

1 Inanna's *Huluppu* Tree, Pomegranates, and Sexual Power 5

2 Pomegranate as Eve's Apple 33

3 Mandrake, the Love Apple, and the World's Religions 55

4 Artemisia, the "Mother Herb" 79

5 The Chaste Tree 113

6 Hermes, Herbs, Elixirs, and Witches 129

Notes 149

Bibliography 185

Index 207

Acknowledgments

I acknowledge the assistance of two colleagues who have read parts or the entirety of this manuscript; each rendered invaluable, expert suggestions and corrections. John Scarborough (University of Wisconsin) and Anne Van Arsdale (University of New Mexico) are extraordinary colleagues whose love of history generously extends to helping others uncover its lessons. Francis Brévart kindly shared with me some of his unpublished researches on wonder drugs, now in print. Kaye Hughes read the manuscript with an editor's eye and two of my students, Robert West and Christopher Musto assisted me with checking citations. I alone am responsible for any remaining errors. Palgrave's editors were enablers in the best sense of the word.

ABBREVIATIONS

ABD	*Anchor Bible Dictionary*, 6 vols. David Noel Freedman et al. eds. New York: Doubleday, c. 1992.
ANET	*Ancient Near Eastern Texts Related to the Old Testament.* James B Pritchard, ed. 3rd ed. with suppl. Princeton, NJ: Princeton University Press, 1969.
AOS	*American Oriental Series.* Chicago-Glückstadt, 1956.
CAD	*The Assyrian Dictionary of the Oriental Institute of the University of Chicago*, 15 vols. Chicago: Oriental Institute, 1956–.
CH-1	*Corpus Hermetica, Hermetica: The Greek Corpus Hermeticum and the Latin Asclepius in a New English Translation, with notes and introduction.* Brian P. Copenhaver. Cambridge: Cambridge University Press, 1992.
CMG	*Corpus Medicorum Graecorum.* Berlin: 1927–.
CML	*Corpus Medicorum Latinorum.* Leipzig: Teuber, 1915–.
DAB	*Dictionary of Assyrian Botany.* R. Campbell Thompson. London: British Academy, 1949.
H	*Hermetica: The Ancient Greek and Latin Writings which Contain Religious or Philosophic Teachings Ascribed to Hermes Trismegistus.* Walter Scott, trans., in 4 vols. Boston: Shambhala, 1985.
MTU	*Die babylonische-assyrische Medizin in Texten und Unterschungen.* Franz Köcher, Robert D. Biggs, and Marten Stol, eds. 7 vols. Berlin: de Gruyter, 1963–2005.
OED	*Oxford English Dictionary.* Electronic edition.
PL	*Patrologiae cursus completus. Series Latina.* J. P. Migne, ed. Paris: Migne, 44–1882., 221 vols.
RAA	*Revue d'Assyriologie et d'Archéologie Orientale*, 1884–.
RA	*Reallexikon der Assyriologie.* 12 vols. Berlin, New York; W. de Gruyter, 1932–.
ZA	*Zeitschrift für Assyriologie und verwandte Gebiete.* Berlin: de Gruyter, 1886–.

Introduction

I know when the good goddess gave humans the knowledge of how to plant seeds and begin agriculture. I know because my professor in Chapel Hill told us back in the early 1960s. According to Professor Wallace Caldwell, it went like this: It was a bright, beautiful, crisp day in late March in the year—or about it—11,304 Before Our Common Era. The place was the hills of Kurdistan. The person who communicated with the goddess was Ugla. Gods and goddesses were mysterious forces that protected and helped people, but they could also be capricious and potentially mean. The important thing was to know what pleased them and stay on their good side.

Ugla had spent much of the day gathering seeds of a wild, winter, nutritious grass to go with meat that the men were expected to bring home from their hunt. Johnny, a prepubescent boy, was to help her, as he was not old enough to hunt with the men, although he thought himself ready. Little by little, a stalk here, one there, Ugla gathered the wild rye that she placed seed by seed in a wicker basket that her tribe had learned to make more than 40,000 years before her day. Johnny was more of a hindrance than help. Ugla and Johnny went back in the campsite, where she was attending to the fire, when something marvelous happened. Ugla told Johnny to fetch the basket so that she could crush the seed on a flat rock with a hand-held rock. Her bread would be crude, this being a time before the discovery of yeast, but I have had bread in places in the world that must have been close to that of Ugla's culinary art.

In a series of utterances, what she spoke had a subject ("Johnny"), a verb ("bring"), indirect object ("to me"), and a direct object ("the basket"). Language in its complexity predated history. Alas, clumsy Johnny tripped and spilled the basket. Ugla spanked him, hard, and she did so with language her kin judged to be profane if not obscene. Furious Ugla and crying Johnny picked up what seeds they could recover, but the emotion and anger of the moment caused some loss of seed.

Five months later, Ugla's tribe returned to the same campsite, since her tribe followed the annual animal migratory herds as they moved between winter and summer pasture lands. What did she see? She saw grain growing in greater density than she or anybody had ever seen and at the very spot where she spanked Johnny. Quickly she reasoned that the Force was pleased by what had taken place. She regarded the Force to be a goddess, because women bring forth life. Ugla sought to duplicate the scene. In the same basket she gathered the seed grain, much easier now that it was in the same place where Johnny fell. She instructed a recalcitrant, whiny Johnny to spill, purposefully, a basket of the grain. She called him with the same words as before, he spilled it, and she spanked him uttering the same, nasty words. By golly, the good goddess was again truly pleased. The following year, Ugla performed the same ritual, but this time she had Johnny to spread out the spilled seed. And she was able to do this in other campsites, so that there was a small field of grain awaiting their arrival each time. After a period, her clan reached a decision perhaps as late as a generation or two after Ugla, who had duly instructed her female successor about the means to communicate with the goddess. Knowing the ritual that pleased the goddess who caused grain to grow, they reasoned that it was easier to stay in one place: plants are more reliable and less perishable than meat. Agriculture had begun because humans knew how to please the goddess. The animal herds would come by twice annually and, between the migrations, the humans could supplement their plant food with meat from small game and fish. Although from our perspective the agricultural revolution was a monumental technological advance in human history, Ugla's people thought it was a divine blessing because they had learned to communicate with the goddess and how to please her so that she would continue giving her blessings.

How did Professor Caldwell know so much about an event approximately 7,000 years before writing and more than 13,000 years before his time? Professor Caldwell had slitters of fact and from them spun his learned, albeit conjectural, tale. In the flat river areas south of Kurdistan before the grain could be sown in the spring, a priestess spanked a prepubescent boy; the words she used were so archaic that no one knew what they meant. The words were magic, because they forced the goddess to grant them fertile fields. The goddess was more esteemed than any other force and she was given a name. There was a gap of over six millennia between Ugla and the rituals used in Uruk and early Sumer. How do we know that Ugla was in Kurdistan? We found some of the earliest agricultural villages there. Weather?—well, that was conjecture, but probably at that time of the year it was clear and crisp. I suspect that Professor Caldwell made up the names of Ugla and Johnny.

By the time humans could write, first in Mesopotamia, a little later in Egypt, the limited number of edible plants that humans could metabolize in their digestive tract was already known. The discovery of most food plants can be traced to prehistoric times, with only a very limited number discovered in a historic period. The same, I am led to believe, happened to the medicinal uses of plants and other medicines from animals and minerals. The discovery of medicinal uses began before humans were humans; many of us believe in the evidence of primate and other animal medicinal uses of plants. Animal pharmacognosy is the term we now call animal effective use of medicinal plants. A plant was ingested and found to be a laxative; another, placed on a wound, eased the pain, and promoted healing. Animals learned and so did humans who followed. One taught the other as mother teaches daughter and father, son. Unfortunately in the last millennium, we have lost much of what humans learned. In this study some of what we learned and how it changed history will be explored and, equally important, why we lost what we learned.

The forthcoming chapters will name some of the most important herbal medicines known before history and relate how each plant is intertwined historically so as to alter the course of events. I claim originality, but my stories are built on the work of outstanding scholars (Sumerologists, Egyptologists, Assyriologists, biblical scholars, classicists, and medievalists) and excellent scientists (botanists, chemists, biochemists, pharmacologists, microbiologists, medical anthropologists, reproductive physiologists, and veterinary and human clinical physicians). These scholars' and scientists' talents enable us to understand the historical record through clever chemistry and sharp clinical observations.

A score or more years ago, I taught a component in a course on medicinal plants in near-by Duke University's Medical School. My lectures were those of an historian; in one lecture I told the physicians-to-be about the transmission of the manuscripts of Dioscorides, the prominent pharmacist whose books were consulted from the first through the sixteenth century. While showing one slide of a beautiful medieval herbal, I told them what the plant was alleged to have cured. A student asked: "Well, did it work?" Embarrassed, I replied: historians don't answer that question. At this reply, he put down his pen with a thud and took no more notes—neither did others in the class. I realized that it was a reasonable question that I had to confront if I had anything to teach them.

In the years that followed, I concluded that most of the drugs of ancient West Asia and Egypt were more a linguistic problem of identification than a medical determination. Not until we get to the works attributed to Hippocrates, the Father of Medicine, can we reasonable identify most of the plants and other drugs. I identified the plants that the Hippocratic writers used, some 257, and found that 89.5 percent of what they used as drugs are found in modern guides to pharmacy and pharmacognosy.[1] Of the twenty-seven exceptions not found in our pharmacy guides, which the Greeks named as drugs, we consider most to be foods. Two examples of the twenty-seven exceptions: spinach was a drug to the ancient Greeks, but we would not place it in our pharmacopoeia. Why not? If one has dietary iron deficiency anemia (many of us do, myself included), spinach ingested acts like a drug. Another example was water, which no pharmaceutical guide today would call a drug. Why not? Hydration is extremely important for health, and dehydration is treated with water. The leading cause of infant mortality is diarrhea, whose address is hydration. In short, what the Greeks considered to be drugs, we know through our usage to be drugs. This study does not necessarily mean that the ancients used drugs precisely as we do. For example, the Greeks used cannabis to stimulate lactation. To the extent that we use cannabis as a drug, our medical purpose is different. Reasonably, what the ancients knew to be drugs we use as drugs. We use many terms now for herbal medicines—alternative, complementary, natural, among others. Scientists usually call them natural-product drugs, but, in truth, they are products of human learning. In our disregard for the past, we have neglected to learn what our ancient ancestors knew and used to their benefit.

As I researched the history of drugs over these many decades, I realized that certain drugs were associated with goddesses, and the realization led to this book. Caldwell's story about Ugla provides insight about the legends we encounter in these chapters. The stories begin when we have records, artifacts mostly, and legends later committed to writing, but, in time, a happy story becomes sad. In looking for elixir and miracle drugs, Western society associated the goddesses' special drugs with evil intent and witchcraft. What humans gained in knowledge was mostly lost.

CHAPTER 1

INANNA'S *HULUPPU* TREE, POMEGRANATES, AND SEXUAL POWER

Some learned thieves caused a catastrophe in Iraq on April 10, 2003. They robbed from and, in other cases, destroyed their and our ancient heritage from the Fertile Crescent. They broke into the archaeological museum in Baghdad, a city "liberated" by American armed forces. The first thieves knew what they were taking, namely the most valuable and transportable artifacts, but, seeing what was taking place, other opportunistic looters joined in the fray by taking and, frequently, destroying ancient artifacts with mallets and clubs. Professor Michalowski of Michigan did not exaggerate when he wrote that this was "a tragedy that has no parallel in world history; it is as if the Uffizi, the Louvre, or all the museums of Washington DC had been wiped out in one fell swoop." [1] In three days of blatant robbery, the museum lost approximately 15,000 objects; among the treasures lost was a vase that contained key clues about plants, medicine, ritual, and religion. Who were these thieves? We do not know, but an Italian helicopter captured pictures of the looting from a predictably safe height. On April 14, the day after the museum looting ended, the museum's director saw the devastation and lamented: "It's gone, and it's lost. It's too late. It's no use. It's no use."[2]

However learned the thieves were, they did not know that one of the earliest artifacts there was a vase from ancient predynasty Uruk, containing a relief that reveals a clue so profound that it causes us to reexamine our understanding of medical botany, family history, ritual, and religion. They did not know; this book proposes to unlock the meaning of the vase. The vase is named after the city of Uruk where it was found near the modern Iraqi city of Warka. (Sometimes the vase is called by the Arabic word Warka, whereas Uruk is called Erech in the Bible.) The so-called Uruk period (*ca.* 4000 to 3100 BCE) is the predynastic period immediately before the Sumerian culture. The vase contains the clues that connect the legend about Inanna, the Sumerian fertility goddess with the legend about the *huluppu* tree and pomegranates. Alas, these thieves seemingly unappreciative of history shattered the famous vase. In chapters one and two, I propose to disclose that the Uruk vase displays a profound insight about the biblical Genesis story of creation. The meaning of Genesis, the Garden of Eden, and the fall is made clearer by this shattered vase that is dated between 3300 and 3100 BCE, a date more than 2,000 years before Genesis was written and 2,500 years before the comparable passage about Adam, Eve, and the Garden of Eden. The Uruk vase was manufactured and artistically decorated 5,000 years before our time.

Before we look at what was on the vase, a few words about how its pieces came back to the Baghdad Museum. The looting, having been allowed—if that is not too fine a word for the circumstances—was the responsibility of the Americans who, together with the museum staff, had to pick up what pieces there were. Art and archaeological dealers worldwide were alerted to look for salesmen bearing the ancient artifacts. Within weeks of the destruction, Col. Mathew Bogdanos (U.S. Marines), a classics major with previous law-enforcement experience, volunteered for the assignment to reconstruct the museum as best he could.[3] He declared an amnesty for those returning its treasures and "no questions" would be asked; he even served the customary tea to those returning artifacts. Many pieces were returned, even some taken from the museum shop; among them was the Uruk vase in the back of a pick-up truck. It had been broken, with pieces from its art panels missing. "No questions" meant no information of who took it; we may conjecture the "learned" thieves may have taken it, but, inasmuch as the vase was heavy alabaster more than a meter high, it probably was broken in transit. Doubtlessly realizing that an attempt to sell its pieces was too risky, the thieves (or their surrogates) handed it over to Col. Bogdanos's men and either no question was asked or it was not reported.[4]

Inanna's Vase from Uruk

I saw the Uruk vase, depicting Inanna, goddess of love, when I visited Baghdad in August 1973. The temperature was 50° Celsius (slightly more than 120° F.) in a day before much in the way of air conditioning had reached Baghdad. I think of Gertrude Bell, who died at the age of fifty-eight in Baghdad in the same month during a heat wave in 1926. She was a wealthy British subject who stopped in Baghdad on the Orient Express train to Iran in 1892 and seldom left Iraq as she developed a passion for the art and history of ancient Iraq. Her collection formed the basis of the National Archaeological Museum; indeed, she designed the building and she restored artifacts. The Uruk vase had been found near the goddess's temple in a pit of votive offerings, where an accumulation of offerings was collected by priestesses and placed in one pit. Therefore, the vase was not found in situ, hence a more precise dating was prevented.[5] The date of approximately 3,000 BCE refers to the pit, not to the date of the objects therein. Little did I realize in the heat of August 1973 that the vase contained information tying the pomegranate to women's lives, powers, and rights. How a simple plant fruit could change the way we live is a story yet to be told in modern times. Embedded in the story also is our perceived relationship to ethereal powers.

Let us begin with the Uruk vase (see figure 1.1). The vase's base holds the key to Inanna's power and, with it, women's rights, but let us start at the top of the vase and move down, frame by frame. The top images were partially broken in

Figure 1.1 Uruk Vase (ca. 3100 BCE) before 2002 looting; top panel, Inanna standing before reed doors of storehouse temple; second panel, processional of offerings; third, rams for sacrifice; bottom, alternating grain and pomegranates.

Source: Iraq Museum, Baghdad, Photo credit: Bildarchiv Preussischer Kulturbesitz / Art Resource, NY.

antiquity. Among the reasons we know that a goddess is indicated is the distinctive headgear with a horn, always the sign of divinity.[6] Either the goddess herself or her representative on earth—here an arguable theological point that I shall side-step—stands before her temple (symbolized by reed poles). Before her is a naked servant offering her a basket of fruits while to his back stands a clothed high priest with a sash around his waist held by another servant (see figure 1.1). This person is variously identified as the En, the priest-king of Uruk, or possibly Dumuzi, a god and Inanna's consort in legend. Quite possibly some or many of those who witnessed the ancient ritual would have conflated the two, En and Dumuzi, in their minds with implicit encouragement from the temple's priestesses. Behind Inanna are sacrificial ram, baskets of fruit, and two votive vases similar to the one we have—or had before the 2003 looting.

The panel below shows baskets of fruits, vegetables, and other agricultural products, which would be the first produce as an offering to the goddess who made the land productive. The bearers are slightly plump, to us possibly on their way to a Weight-Watchers meeting, each naked and shaved of all body hair (just as were early Egyptian priests). The next to the last panel shows male and female rams, alternately in a march probably for sacrifice and feasting.

Finally, the base and foundation (in more ways than one) underlie the story above. Geometrically the pictures alternate stalks of grain and a limb bearing three pomegranates (see figure 1.2). The grain is easily understood because this practice has continued for these 6,000 years. At the end of the marriage ceremony, the bride and groom are showered with grain as a symbol of fertility. That is, after all, the realm of Inanna, who is often referred to as the Queen of Heaven and Earth.[7] From antiquity to the present, the ceremony of throwing grain on the bride and groom has persisted. What varied was the type of grain, such as rye, barley, wheat, and, since the eighteenth century in America, rice, and still more recently, birdseed.[8] Indeed, much of the modern marriage ceremony goes back to antiquity: father giving away the bride; bride and groom exchanging vows in front of priest with witnesses; best man; exchange of rings; the ceremonial eating of a wedding cake by bride and groom. The list can continue but without purpose to explain why the nuptial pair is showered with grain and given pomegranates.

Equal to grain in the depiction is the pomegranate. If the throwing of grain was an important clue to Inanna, women, and marriage, what happened to the pomegranate, I asked in my class? In the class was an Iranian student who said that in her country on a wedding day, the mother of the bride gave her daughter a pomegranate. Why, I asked the student. "Because it is traditional," she replied. Why was there continuity in the pomegranate and grain on the vase with present Iranian wedding practices, I asked, and now I seek an answer in this book: simply Googling "pomegranate," "Iran," and "wedding," confirms the current practice of pomegranates associated with weddings (and Greek nuptials as well).[9] Heretofore, scholars advance no reason for the pomegranates on the Uruk vase, but there is a reason, and it is the key to understanding Inanna's power. The grain symbolized fertility and the pomegranate symbolized fertility's control, namely contraception. The symmetry is compelling: grain for fertility; pomegranate for its control, two powers that women would normally desire. Before I offer scientific proof for pomegranates' contraceptive

Figure 1.2 Lower panel of Uruk vase showing alternate stalks of grain and limbs each bearing three pomegranates.

Source: Iraq Museum, Baghdad, Photo credit: Bildarchiv Preussischer Kulturbesitz / Art Resource, NY.

action, we will first probe deeper into the vase's story and relate Inanna's story as written in cuneiform documents. Denise Schmandt-Besserat, a modern scholar of the period when art evolved into cuneiform writing, argues that around 3500 BCE a "fundamental change" occurred in the use of art for communication.[10] Art became linear and was used to evoke a story. The Uruk vase is a prime example of a complex story related through pictures. After all, in today's world, from our childhood days, we know how vividly pictures in comic strips and books told stories when the words had no meaning to our juvenile eyes.

Inanna, Queen of Earth and Heaven

Possibly it is Dumuzi, the one with the sash, who stands before her, rather than, as some scholars postulate, the high priest-king, En, the title of Uruk's secular ruler (see figures 1.1 and 1.2). The fact that his height is the same as that of Inanna (while the servant bearing gifts is smaller) indicates a special, divine status. The top frame of the vase, it is thought, represents Inanna's sacred marriage to Dumuzi, her consort. If we can read backward from the next period of the Sumerians when evidence is more substantial, the common marriage ceremony began with the groom bearing gifts of food and going to the home of his beloved. If the family lets him in, accepts the gift, and allows him to spend the night, the marriage is consummated. The next day there was a feast to which many guests were invited. At Inanna's back the temple door is a storehouse of food.[11] Conjecturally as depicted, the ritual is the New Year celebration that acknowledges Inanna's fertility in bringing fertility to the land and its animals that nourish and serve the people of Uruk.

Inanna was the beloved Queen of Heaven and Earth and responsible for life and growth on earth, the fertility of plants and animals as well as humans. From hymns and rituals, we know that Dumuzi symbolized fertility and the renewal of life in the spring.[12] Forces of the Underworld killed Dumuzi, Inanna's groom, and Inanna would search for him in the Underworld, a story we shall shortly relate. Uruk religion asserted that there was a marriage between heaven and earth and life and death, themes carried into Sumerian religious beliefs.[13]

Inanna came to Uruk from the neighboring city of Eridu and tricked Enki, the masculine fertility deity, a phallic representative of fertility, into giving her powers. Subsequently she became Uruk's central deity. Modern scholars explain the story on the basis that Uruk achieved hegemony over Eridu. Inanna presided over the temple where agricultural production was stored. Indeed, the earliest proto-writings were marks on clay to indicate the types of commodities (usually grain and animals) and the measure of quantities that individuals stored in or delivered to the temple, such as baskets of rye, or rams.[14] Inanna's human representative, a woman to be sure, had a consort (husband?), called an *En*, who managed the temple's estates.

Modern historians cite Inanna as an indication of a matriarchal society that was transformed into a patriarchal society when, in time, the *En* functioned as a king and protector of the city, thereby reducing the Lady's power.[15] To interpret the phenomenon the spiritual function represented by Inanna became subservient to the function of protection. Samuel Noah Kramer, who worked for decades on the texts surrounding Inanna and Dumuzi, believes that Dumuzi had once been a "mortal... ruler" whose heroic status continued and elevated him to the status of a deity, as seen in the annual New Year celebration of the marriage of Dumuzi to Inanna.[16] The Uruk vase, when interpreted in light of Sumerian myth stories preserved on cuneiform tablets, informs us about Inanna, Dumuzi (or his surrogate), and some rituals, as we read the pictures as one would a comic book without writing.

It is from the writing of one particular matrix of stories that we gain our greatest clue about Inanna, pomegranates, contraception, and what the gods thought people should know. Or, better: what people thought the gods thought they should know. During the third millennium BCE, Inanna, the Sumerian goddess of love, was worshipped in temples.[17] Who was she and why did she develop a power that made her beloved?

Records of Inanna's Story

The written story dates from at least the Early Sumerian Dynastic I period (*ca.* 2900–2750 BCE), which followed Uruk's predynastic time (3500–3100 BCE), but, being based on oral tradition, likely goes back to Uruk, whose city deity was Inanna. Sizeable fragments of two major Sumerian stories have, happily, survived: one is the

Gilgamesh Epic, known to many inasmuch as the story is taught in many schools today; the others are two related stories: "Inanna and the *Huluppu*" and "Inanna's Descent into the Underworld." The last two titles are what scholars have reconstructed from fragments. The University of Pennsylvania conducted four archaeological expeditions between 1899 and 1900 in the Nippur region of Iraq, then governed by the Ottoman Empire. The expedition uncovered a number of tablets in cuneiform. By agreement, artifacts were divided between the Istanbul Museum of the Ancient Orient and the Museum of the University of Pennsylvania. It takes a great deal of time to read a tablet, especially one in poor condition. Scholars skilled in cuneiform reconstructed these tablets in a logical order and translated the texts in a little more than half a century. The analogy would be like working on the most complex jigsaw puzzle with half the pieces in places half the world away. The reconstruction of the Inanna story was not to be accomplished until scholars, particularly Samuel Noah Kramer, recognized that a key tablet was in the Yale Babylonian Collection and allowing him to piece together a reasonable reconstruction of the early tale.[18] Some scholars (among them Kramer) regard the Descent and *Huluppu* stories to be later and inferior compositions, compared with the Epic of Gilgamesh. A number of recent scholars consider these stories as part of a unified story and composition.[19] In unfolding the story and its meaning, let us begin with Inanna's Descent into the Underworld, where she gains her power from the *huluppu* tree, and, in the end, she returns to become a major if not the preeminent goddess of heaven.

INANNA'S DESCENT INTO THE UNDERWORLD

The "Descent of Inanna" to the Underworld tells the story of how Inanna gave up earthly and celestial glory and her family:

> My Lady abandoned heaven and earth to descend to the Underworld.
> Inanna abandoned heaven and earth to descend to the Underworld.
> She abandoned her office of holy priestess to descend to the Underworld.
> In Uruk she abandoned her temple to descend to the Underworld.[20]

Diane Wolkstein notes that the theme is a familiar one: separation from family; progressive regression to a neonatal state; death after suffering; rebirth; and, finally, killing of one person to replace another.[21] The gods and goddesses in all the old Mesopotamian texts are not always wise, just, and powerful, but they act like humans—or some humans—: willful, petty, vengeful, and selfish.[22] In embarking on her self-directed journey, Inanna's previous accomplishments in life were removed and she was "stripped naked, with nothing remaining but her will to be reborn."[23]

When Inanna reached the gates of the Underworld, Neti, the gate's guardian, asks her:

> Why has your heart led you on the road
> From which no traveler returns?[24]

She gave what appears to be a contrived story: to attend the funeral rites of her sister's husband and to do service to her sister, but it seems clear that her underlying motive was a journey of discovery and sexual satisfaction. Permitted to enter, Inanna finds seven gates at each of which she must give up things of value, such as jewelry or clothing, beginning with her crown. At one gate she was deprived of her beloved Dumuzi.[25] When she passes the seventh gate, she is left "naked and bowed low."[26] At the seventh gate, Ereshkigal (lit. "great lady under earth") bids the naked Inanna:

> Be silent, Inanna! It befits the ways of Hades;
> raise not your voice, Inanna, against Hades' customs:
> crouched and stripped bare, man comes to me.[27]

In another fragment of her story (called "Inanna and the God of Wisdom"), she says that her joy is her vulva, her possession in nakedness:

> When she leaned against the apple tree, her vulva was wondrous to behold.
> Rejoicing at her wondrous vulva, the young woman Inanna applauded herself.[28]

Other Sumerian texts refer to Inanna's silver *urû* (vulva).[29] She wore a pectoral named "Man-come."[30] In the Underworld, Inanna eats clay, drinks dirty water, and has no compassion or feeling for others. Although still existing as an individual, Inanna has died. As retold by the Sumerian poets of the third millennium BCE, Inanna sought fulfillment of her feminine powers by a descent into the Underworld, first as an innocent virgin, returning as a mature woman who knew sorrow and had met death. Whereas her family was furious with her for having gone to the Underworld for no good reason, Inanna's mother's father, Enki, the God of Wisdom, understood his granddaughter, and joined her in the Underworld, because he also understood that Inanna was vital for all lands. As Wolkstein observes, "God of Wisdom" for Enki is a metaphysical translation. The Sumerian word can mean "God of the Waters," and "waters" can also mean semen.[31] The sexual connotation, however, was probably not primarily intended to be "wisdom." At this point, the reader should note for future reference that Inanna met the God of Wisdom in the Underworld when she achieved her power through the *huluppu*, which facilitated her "rebirth." She desires to return to the Earth and Upperworld, but "no one ascends from the Underworld." Inanna is different, the first to be reborn and return from death, but even she must pay a price. She must replace herself, and that she does with her sister Ereshkigal. Her substitute, Dumuzi undertook the cyclical visits to Ereshkigal in the Underworld each year on Inanna's behalf. It is tempting to infer that the cycle of going to and returning from down-under was an explanation for the cycle of spring-summer and fall-winter, as later explicitly stated in Greek mythology; the inference is only vaguely implied in the Uruk-Sumerian story. So much is told in the tablets known as Inanna's Descent into the Underworld. Her power came while in the Underworld, as told in "Inanna and the *Huluppu*," which begins with Ereshkigal, the Queen of the Underworld. If you are confused by the chronology of "events" of how Ereshkigal was, on one hand, the substitute hostage for Inanna and then appears as the Queen of that hellish domain, keep in mind two things: one, we are dealing with tablet fragments (albeit some large) of stories, and, second, religion is not always clearly rational to us mortals.

INANNA AND THE *HULUPPU* TREE

Ereshkigal reacted to Inanna and Enki's presence with a storm, hurling wind and hail stones, and threatening to devour them. Wolkstein interprets the story to mean the struggle between male and female, but Inanna is given a secret weapon that increases her power, attractiveness, and sexuality. Some modern interpretations conclude that Inanna and Ereshkigal were once one goddess, embedded as two sisters; Inanna became the beloved goddess of the World and Upperworld, whereas Ereshkigal presided over the Underworld. Some of these interpreters see a male dominance in telling and retelling, whereby it took a male to rescue Inanna: Enki, on one hand, and Gilgamesh (as we shall shortly see) on the other.[32]

Inanna's power was acquired when she gained the *huluppu* tree, in her sojourn in the Underworld. The poem reads in English translation:

> At that time, a tree, a single tree, a *huluppu*-tree
> Was planted by the banks of the Euphrates.
> The tree was nurtured by the waters of the Euphrates.
> The whirling South Wind arose, pulling at its roots
> And ripping at its branches
> Until the waters of the Euphrates carried it away.[33]

An unnamed woman, probably meaning generic woman, who feared the gods, "plucked the tree from the river" and gave it to Inanna. The generic woman is an important clue, because the tree possessed a power for all women. Inanna spoke:

> I shall bring this tree to Uruk.
> I shall plant this tree in my holy garden.
> Inanna cared for the tree with her hand.
> She settled the earth around the tree with her foot.

She wondered:

> How long will it be until I have a shining
> throne to sit upon?
> How long will it be until I have a shining bed
> to lie upon?'
> The years passed: five years, then ten years.
> The tree grew thick,
> But its bark did not split.[34]

Inanna wanted a throne and a bed, as Wolkstein surmised, she "wanted the end result—her rule and womanhood."[35] Her happiness was in exercising her sexual powers but, alas, three powers intervened to thwart her power and end her happiness. The poem relates how, much to Inanna's chagrin, a serpent "who could not be charmed" built a nest in the *huluppu* tree's roots, how "the dark maid" Lilith built her home in the tree's trunk, and how the *anzû* bird raised its young in the tree's branches.

> The young woman who loved to laugh wept.
> How Inanna wept!
> (Yet they would not leave my tree.)[36]

The *huluppu* tree would bear no fruit because of the snake, Lilith, and the *anzû* bird. In many lines of the poem, Inanna calls upon her brother, Utu, the sun god, and her brother, Ninurta, the valiant warrior, to remove the serpent, the a*nzû* bird, and the dark maid Lilith from her tree, all to no avail. In desperation, she turns to Gilgamesh, "the hero of Uruk, [who] stood by Inanna." He fastened his armor, well described by the poet, and "entered Inanna's holy garden."

> Gilgamesh struck the serpent who could not be charmed.
> The *anzû*-bird flew with his young to the mountains;
> And Lilith smashed her home and fled to the wild,
> uninhabited places,
> Gilgamesh then loosened the roots of the *huluppu*-
> tree;
> And the sons of the city, who accompanied him, cut off
> the branches.[37]

From the tree's trunk Gilgamesh carved a throne and a bed for the goddess, Inanna. From the tree's roots and crown Inanna fashioned a *pukku* and a *mikku* for her servant, valiant Gilgamesh. The *pukku* is likely a ball or puck and the *mikku* a stick, implements for athletic contests.[38] A related passage from a cultic lament (*Uruammairrabi*) speaks of boys playing with *pukkus* while girls skip rope.[39] The meaning of the *huluppu* tree is central to this goddess' ability to be reborn. Already she had discovered her sexuality, so for fulfillment she found the tree's presence in her garden necessary, and upon it her throne and bed would be built. In light of our interpretation of the tree as an antifertility agent, the tree's occupants can be reinterpreted.

DARK MAID LILITH

The Sumerian word, *ki.sikil.lí.lá.ke* (or, *ki.sikil.líl.lá*), was translated by Kramer as Lilith found in a Sumerian Gilgamesh fragment. In Assyrian the word is *lilītu*, feminine of *lilû*, a female demon and enemy of pregnant women who should not come near a baby.[40] The Talmud has four references to her, and other Hebrew sources sketch out details lacking from earlier sources.[41] She was known as Adam's first wife whom Adam expelled because she sought to be his equal. In time she became the first witch, in the words of the legend: "She takes her revenge by injuring babies—baby boys during the first night of their life, while baby girls are exposed to her wicked designs until they are twenty days old."[42]

In the Middle Ages she was known as the first witch. Both in antiquity and the Middle Ages, she is depicted as endangering babies by sucking their blood, eating their marrow, and consuming their flesh. In particular she attacked during childbirth, for which occasions ancient women had amulets to protect against Lilith. Men were not unprotected from her wrath, but they were safe if they slept alone, thus a strong sexual implication goes along with baby-snatching.[43]

Kramer's identification of the term in the Inanna story and Lilith of later lore is challenged by Sergio Ribichini as being an unmerited assumption. In its place he proposes that the term "Lilith" designated the opposite of fertility, namely infertility and sterility, with the imagery of living in the tree's cavity.[44] Whatever the interpretation of the linguistic evidence, the "dark maid Lilith" certainly lends credence to the interpretation of a demon or spirit who captures pre- and postnatal babies.

THE *ANZÛ* BIRD AS THE WATER-PREVENTION MONSTER

The meaning of the *Anzû* bird is more difficult. The *anzû* bird is not identified by modern scholarship, although representations in artifacts somewhat resemble an eagle.[45] But another text (*Enmerkar and the Lord of Arrata*) clearly differentiates the *anzû* from an eagle:

> This great mountain range is a *meš* tree grown high to the sky; its roots form a net, and its branches are a snare.
> It may be a sparrow but it has the talons of an *anzû* bird or of an eagle.[46]

The text known as the "Epic of Anzû" connects the birth of the *anzû* bird with a mountain called Šaršar,[47] but this particle of knowledge does not help us to connect with the dot in the Inanna legend about the *huluppu* tree. There is possibly one link that connects the *anzû* with the *huluppu*; this occurs in a series of tablets on the Sumerian and Akkadian hero Ninurta, god of agriculture and water. Inanna is not mentioned, but variously the Anzû is referenced as a bird and a monster that once was good but turned evil and blocked water. In translation, Ninurta "brought forth the *Anzu*d [Anzû] bird from the *halub-haran* tree."[48] *Halub-haran* is phonetically close to *huluppu.* Elements of other legends are intermixed, including *Enuma elish*, the Sumerian creation story involving primordial forces of water, and the Gilgamesh Epic.

Although modern scholars have not identified the *anzû* bird with a specific bird, later history may provide clues that the bird was the owl. Ovid says that the person who betrayed Persephone's eating of the pomegranate seeds was Ascelaphos (see chapter Two on Persephone). For being an informer he was turned into an "ill-omened bird/ *avem sparsum*" with a beak, feathers, and big eyes.[49] The loathsome bird was the slothful screech owl, famed in medieval lore in association with witchcraft. Indeed an old medieval Latin term for witch, *scobas,* meant in Greek screech owl.[50] No direct evidence exists for association of the *anzû* with either the cuckoo or owl, but then there is no evidence for any type of identification. In chapter six, more evidence will be forthcoming to hypothesize the bird as being an owl. The *anzû* bird represents a force that blocks the watering of the *huluppu* tree, thus preventing it from producing its fruit, the pomegranate. The *anzû* formed a nest in the *huluppu's* branches and, when expelled by Gilgamesh, took its young to the mountains.[51]

The Serpent

The reader of the Genesis text must supply interpretation because the serpent's meaning is not explicit. For interpretation, let us look at the tree's occupants: the serpent could be the general evil toward which fertility and its absence can be directed; the dark maid Lilith is the evil purpose for which abortion and contraception can be employed; the *anzû,* the means by which the deed was enacted. Much later (as seen in chapter six) was the serpent the personification or representation of the Devil (or Evil) and the bird was the agent of enactment. A study of cuneiform texts and other ancient Western Asian writings and figurines will show that the serpent was commonly related to creation and fertility. The serpent comes from the realm of Mot, the god of the Underworld and death; the contrast is made between death and life, the "tree of life" and infertility, human fertility, and agricultural fertility.[52] Although much later in time, when St. Paul, the apostle, was bitten by a serpent, the people's reaction was that he was an evil man, a "murderer,... [thus he should] not be allowed to live."[53]

What Is the *Huluppu* and Its Meaning?

Let us return to the Uruk vase that has grain and pomegranates in juxtaposition. As a woman Inanna was fruitful—one of her titles was often "Lady of Fruitfulness and Sexuality"[54]—thus, there was the depiction of grain. Her happiness came with the fruit of the *huluppu* tree. When she went to the Underworld, she sought not sexuality—she already had that—but an unfulfilled desire for happiness that she found in the power imparted by the fruit of the *huluppu.* The fruit was, I propose, the pomegranate, because she could not possess a woman's power and happiness until she could control her fertility by contraception. First, let us look at the evidence for *huluppu* as pomegranate.

Sumerologists speculate on the translation of *huluppu*, the date palm and willow having been suggested.[55] Some scholars, such as Marvin A. Powell, doubt that a secure identification can be made.[56] On the other hand, reference to a tree called the *huluppu* is found in various texts, including Assyrian, thereby demonstrating that the word was not literary imagination. The *huluppu* tree (variant transliterations as *haluppu* and *hiluppu)* grew on a plantation in the Harran region *ca.* 700 BCE.[57] Certainly the date palm (which will be discussed in chapter two) is the most common tree in Mesopotamia, and there is ancillary art work that associates the date palm with Inanna.

Other examples of the *huluppu* tree occur in various tablets that state that chairs, pegs for a chariot, tables, and bowls are made from it. Everyday physical items should rule out the hypothesis that the tree is legendary. The possibility of its being an oak tree is proposed based on its use for carpentry objects and the fact that another word for oak is not identified in the texts.[58] Probably oak grew in ancient Mesopotamia.[59]

An Old Babylonian text refers to *ha-lu-b'a* as a term for furniture, and it is proposed as meaning oak.[60] Still the *huluppu* wood is never used for the construction of boats,[61] a fact that suggests that oak is highly unlikely. Moreover the objects made from the tree are small, allowing the possibility of a small tree or large shrub. A text of a religious ritual casts doubt on whether it was a date palm, because date palm is specified in addition to *huluppu*: "by means of seven date palm [branches], seven *huluppu* [branches] and seven. . . ."[62]

INANNA'S NAME

Thorkild Jacobsen proposed that Inanna is from the name *Nin.an.na* meaning "Lady of the date cluster."[63] Helgard Balz-Cochois advanced a new interpretation of Inanna's journey and her name. She takes a clue from the epitaph applied to her, *nu.gig*, usually translated as "courtesan," "prostitute," or "hierodule," for those more classical-minded. *NU* means "sexual" and *GIG* "sickness." Thus Inanna is *nu.gig.an.na*, implying menstruation. The meaning is not fertility per se, as is the usual association of Inanna, "the fertility goddess," but the menstruating goddess, one who can be fertile or nonfertile.[64] The fact that the pomegranate produces sex hormones could, depending upon the circumstances of the patient or recipient, the dosage, and the timing, be a substance that either contravenes fertility or promotes it.

POMEGRANATE AS *HULUPPU* TREE

The identification of the *huluppu* tree as a pomegranate has not been considered. The pomegranate is a small tree or large shrub (approx. 15 feet high) and could be the source of wood for chairs, tablets, pegs and bowls, just as the rhododendron shrub (approx. the size of a pomegranate tree) in North Carolina serves as the material for furniture. An Old Babylonian text from Ur mentions a pomegranate *giš.nu.u'r.ma.la'l.da'r*) beam three meters long.[65] The pomegranate grew in southern Babylonian orchards together with date palms and apples, all three being used for wood.[66] One Assyrian prescription associates pomegranates (*giš.nu.ur.ma*) with love magic: "If (he is in his seventies and) *hu-qu* (gasping respiration) continually afflicts him (and) he asks for pomegranates (*giš.nu.ur.ma*), he will die."[67] (The preword "*giš*" means "little tree.") The meaning for this warning is too vague for speculation except to say that the pomegranate was associated with love and, in this case, with death for a male in his 70s.[68] The translators say that the context is love magic, but its meaning they place in the appendix of "unsolved puzzles."[69] The existence of an Assyrian term, *giš.nu.ur.ma*, for pomegranate does not impeach the thesis that the *huluppu* is also a pomegranate tree and, in the case of this "unsolved puzzle," the context was sexual. Also recipes using pomegranate, using the same cuneiform, were given orally for difficulty in urination and another urinary problem for which translation is lost in the text.[70] The legend about Inanna dates back to the predynastic Uruk period before the Sumerians came and much later than the Assyrians (Early Empire period, 933–782 BCE). A different word for the same plant would be expected over a span of 2,000 years from late Uruk and early Sumer to the Assyrians. Because *huluppu* was central to the legend, the Sumerians kept the word; religious texts had some literal meaning, we conjecture, for the Sumerians, as it does for Christians, Jews, and Muslims today.

Pomegranate as Contraceptive

Published in 2003, research on gynecological disorders in Chhattisgarh, India, reported a "birth control measure" in "common practice" by using pomegranate seeds for prevention of conception. Women in Chhattisgarh ground the seeds into a powder, mixed in sesame oil, and placed the mixture in the vagina just after the menstrual cycle.[71] A cuneiform recipe from Aššur refers to the application of pomegranate to wool, which was placed in the vagina.[72] We suspect that the text indicates that it was not to cure a woman's disease, as a translator previously suggested, but for use as a contraceptive pessary (i.e., vaginal suppository). The lining of the uterus will absorb many chemicals, whereas oral administration can often block pathways to absorption.

In 2006, thirty-six scientists, mostly biochemists and medicinal chemists, published in a single volume a number of extensive projects on the pomegranate.[73] The pomegranate (various parts, including seeds, roots, and fruit pulp) contains a number of estrogenic compounds, primarily 17β-estradiol, estroil, and estrone. The compound having the most bioactivity is 17β-estradiol. Women's ovaries and placenta secrete estrone and estradiol. Pomegranates have "a relatively high concentration of *bona fide* naturally occurring estrogens," to quote from one study.[74] Estrones are currently used in contraceptives, among which is the most common oral contraceptive that has estrones and progesterone, another hormone that prevents implantation. Current research on pomegranates as a therapeutic agent is as a replacement or alternative to hormonal therapy for menopausal and postmenopausal disorders.[75] Thus the common contraceptive pill in America is a combination of estrogen (to disrupt ovulation) and progestin (to prevent implantation). Either compound is a contraceptive, but the combination is a double protection to achieve the desired result—the safe prevention of pregnancy. A similar study (2006) determined that pomegranate juice has the three "classic estrogens, estadiol, estroil, and estrone" which had "an effect on vaginal maturation, an indicator of physiological effects" in a search to find a safe postmenopausal therapy. After a week in clinical trials, the study showed "a significant increase in estrone at a nonphysiologic level in postmenopausal women."[76]

Other studies point to the pomegranate's contraceptive effects. In experiments conducted during the 1970s and '80s, female rats fed pomegranate (*Punica granatum* L.) and paired with male rats not so fed had a 72 percent reduction in fertility. Pomegranates fed to guinea pigs resulted in a 100 percent prevention of pregnancy.[77] Forty days after withdrawal from pomegranate as a dietary additive, the fertility of both animal groups returned to normal.[78] Still another experiment with rats had a 50 percent reduction in fertility.[79] In analysis, other parts of the plant (seeds, roots, whole plants) had no appreciable effect on fertility; only the fruit skin around the seed lowered it.[80] One test shows that pomegranate root has substances that inhibit fetal growth by nearly 100 percent.[81]

The same use as that in ancient Mesopotamia and modern India is verified in other ancient medical writings. A pharmacist's *vademecum* (portable reference guide) has this line: "Seed of *haluppu*. Plant for a woman who does not get pregnant." Disregarding the vowel, this is the pomegranate seed, and this tablet, along with the iconographic evidence, confirms the thesis that *huluppu* in the Inanna story was a pomegranate and was used as a contraceptive. Prescriptions for preventing pregnancy and for abortion are found in various Assyrian texts, but with the two inscriptions that we have, the meaning of the plant taken is unclear: "*ú munus nu peš*$_4$ for a woman not to get pregnant"; "*ú šá-,mi arhuš šub-e*... to make fetus drop."[82] When prescriptions in cuneiform tablets point to contraceptives and abortifacients, for example, "to prevent childbirth," Jean Bottéro, a French Assyriologist, dismisses the claims as "magic" or "exorcistic' therapeutics."[83] Bottéro is incorrect: they were real. Magic would, at best, be a placebo or psychological effect, but, if one takes a substance to prevent or end a pregnancy and full term births occur, one knows full well it did not work.

Several recipes in Middle Assyrian tablets prescribe pomegranates (*nu.u'r.ma*) together with cedar resin for tinnitus treatments ("roaring" in the ears), suppurations, and bleeding from the ears. Pomegranate juice has a haemostatic action and is also antibacterial, anti-inflammatory, antiseptic, antiviral, and an expectorant. In translating these texts, modern investigators concluded that the recipes for pomegranate juice would have been effective according to modern science.[84] In the next chapter, Greek medical usages of the pomegranate as a contraceptive will be presented.

Phytochemistry and Pomegranate's Alteration

A general rule in phytochemistry is that chemical compounds may concentrate in one particular part of a plant at various times in the growth periods, but traces may be present throughout. One example is opium, which is concentrated in the poppy ovary during only one stage of growth. Various experiments with pomegranate reflect differing results, which can be attributed to any number of variations. The point is that modern scientific experiments show that the pomegranate inhibits fertility in animal populations. In the Inanna legend we identify the *huluppu* tree with the pomegranate and the pomegranate as a contraceptive. The pomegranate in ancient myths is associated with fertility as we shall see in the next chapter. The pomegranate was a prominent symbol of the goddess of love in ancient West Asia.[85] Religious ritual clearly connects Inanna with the pomegranate. An incantation in cuneiform shows that Inanna was associated with the pomegranate and the apple:

1 [Incan]tation. The beautiful woman has brought forth love.
2 Inanna, who loves apples (*giš.haš.hur*) and pomegranates (*giš.nu.u'r.ma*)[86]
3 Has brought forth potency.
4 Rise! Fall! Love-stone, prove effective for me! Rise!
5 ...Inanna....
6 She has presided over love.[87]

Other entries in this incantation show that the "love-stones/ love-apples" could be either for enhancing fertility or preventing it. [88] A discussion of "love apples" will be seen in chapter three.

Later artifacts connect goddesses of love with the pomegranate.[89] An ivory relief from Aššur near the end of the fourteenth century BCE, now in the Staatliche Museum, Berlin, has a god receiving the streams of water from flasks in the sky and diverting them to the earth. On either side stand pomegranate trees.[90] From the same panel is another scene, this one of the Sumerian god *Sin-Annar* standing between a pomegranate and a palm.[91] A gold plate from Ras Shamra of the same time period shows the god as a bull, representative of Baal, with a border of pomegranates.[92] A wall-hanging from Nimrud datable to the kingship of Shalmeneser III (858–824 BCE) has a similar scene with pomegranates prominently displayed.[93] A magnificent piece of gold-silver alloy jewelry was found in a grave from the seventh century BCE in Kamiros. It shows a naked female figure, likely Ištar or Aphrodite; from it hangs pendants of pomegranates.[94] Two roll cylinders, one from the period of the same king, Shalmeneser III, show the double representation of the king facing a pomegranate tree, each with a pomegranate in hand at the end of a shaft going to a god. The other cylinder comes from Aššur, is dated from the ninth to the seventh centuries, and represents a similar scene, except that the winged spirit of the king that stood behind him in the first cylinder is now at the pomegranate tree.[95] Other reliefs, such as one from the palace of Sargon II (721–705 BCE), represent the king with pomegranates in his left hand.[96] An Assyrian tablet describes a god's knees as being pomegranates (*nu.u'r.ma*).[97] In Mycenean, Etruscan, and Greek art the pomegranate was frequently depicted on pendants, pins, vases, and reliefs, including a number of vases showing Athena carrying a pomegranate.[98]

Friedrich Muthmann proposed that the pomegranate was identified with the fertility deities because it was the symbol of fertility and with kings as the earthly agent for the same.[99] By the Neoassyrian period, the symbol of the pomegranate may well have been for fertility, but we propose that earlier it symbolized fertility control and probably did in the Neoassyrian and later periods. A Neoassyrian incantation demonstrates the point:

> 7 Incantation. If a woman looks upon the penis of a man.
> 8 Either a *giš.hašhur* [apple?] or one *giš.nu.u'r.ma* (pomegranate):
> 9 You shall recite the incantation three times either to an apple or to a pomegranate. Give (the fruit) to the woman (and) have her suck their juices.
> 10 That woman will come to you (and) you can make love to her.[100]

Retold by modern scholars, the incantation was a love-charm. Our point is that it had meaning beyond the power of words and charm. The effect was real according to modern science expectations, given its chemistry. *Hašhur* is usually translated as "apple" but, as we shall see in chapter three, there is doubt about *hašhur's* meaning.[101] In chapter three, another meaning will be proposed for the Sumerian word *hašhur* and Akkadian *hašhuru*.

Before love-making in the Hebrew Song of Songs (8:2) a woman wishes to give to her lover mulled wine and fresh pomegranate juice. Drinking a fruit juice before intercourse appears customary in the Sumerian myth Enki and Ninhursag.[102] In an Old Babylonian Hymn Ištar is described:

> She/Goddess of joy, clothed with love,
> Adorned with fruit, cosmetics and sex-appeal.[103]

In another hymn Ištar is called, "Mistress of fruit," (although possibly in the sense of "fruitfulness").[104] The connection between love-making and fruit of some kind is evident, just as the ritual in the Sacred Marriage between Dumuzi and Inanna makes abundantly clear.

The Sacred Marriage

Many historians consider the topic relief on the Uruk vase as a representation of the Sacred Marriage. As the daughter of Suen and bride (or consort) of Dumuzi, Inanna's conception was ritually reenacted in the Sacred Marriage Rite in temple services.[105] It began with soft music, while townspeople gathered to watch the king, incarnated as or representing Dumuzi, copulate with a beautiful woman, as the goddess Inanna. While to us moderns the ritual was erotic, to them it was a solemn religious ritual to begin each New Year in the hopes of pleasing the god and goddess who would bring bountiful crops, adequate food, peace, and happiness. A terracotta bed on four legs, dating from the Third Millennium BCE, depicts what is thought to be the Sacred Marriage and with words taken from a hymn to Inanna:

> The people of Sumer assemble in the palace...
> So that the New Year's Day, the day of rites,
> May be properly determined,
> And a sleeping place be set up for Inanna...
> The king goes with lifted head to the holy loins,
> Dumuzi goes with lifted head to the holy loins of Inanna.
> He lies down beside her on the bed.[106]

The king was an avatar, a representative as it were, of the good god Dumuzi. His responsibility was to carry out his manly duties and properly satisfy the priestess, who was Inanna's representative. Around the year 2000 BCE, or more than a millennium after the Uruk Vase, a Sumerian poet used words to this effect:

> The rivers flowed fresh water,
> In the field grew rich grain,
> The sea was filled with carp and... fish,
> ...The forests were filled with deer and wild goats,
> The orchards were filled with honey and wine.[107]

For these good things to happen, the king had to maintain a state of erectile function in plain view of his people to carry out his royal duty. Although the Uruk vase does not show, of course, actual copulation, the earliest "sure evidence for ritual copulation between goddess and ruler," assumed to be representing Dumuzi, comes from Ur III during the early dynastic period.[108] One imminent scholar of the period, W. W. Hallo, believes that part of the ritual's purpose was actually to produce an heir as a product of the king and divine descent from the goddess. In Hallo's words: "the sacred-marriage['s] . . . object was to produce a royal heir and to establish his divine descent; . . . [it was a] real act."[109] One must conjecture that behind the scenes events may have occurred when that statistical reality of a single act did not produce the desired results, but royalty and governments have a long tradition in deception.

Inanna and Prostitution

Today's woman takes estrogen to prevent pregnancy or, wishing the same and needing medical intervention, she takes estrogen. Thereby, we should not call Inanna, Ištar, Demeter, and the like, fertility goddesses but fertility-control goddesses.

The story of Inanna's sojourn in the Underworld is, in Balz-Cochois's opinion, that of a premenstrual girl finding her sexuality along with menstruation.[110] I add this interpretation: that is why the possession of the *huluppu* tree was her first endeavor. In Inanna's exhortation when Gilgamesh restored its control to her, she explained,

> My father has given me the *Me*: [*Me* = civilization; domestication]
> . . . He gave me godship.
> . . . He gave me truth.
> . . . He gave me the art of lovemaking.
> He gave me the kissing of the phallus.
> He gave me the art of prostitution.[111]
> The god Enki said to the people of Uruk:
> Inanna brought the *Me*:
> She brought the placing of the garment on the ground.
> She brought allure.
> She brought the art of women.
> She brought the perfect execution of *Me.*[112]

The "perfect execution of *Me*" recalls how the priestess from the "temple of love" transformed Enkidu in the Gilgamesh epic from a wild man who ran with the animals to a civilized person who wore clothes and lived in houses kept by the priestess from the "love temple."[113] Some biblical scholars tied Adam to Enkidu, who, while not the first man, nonetheless was made of clay and lived in a purity of nature with wild animals.[114] In the Gilgamesh Epic, Enkidu was transformed from a wild man who ran with the animals to a civilized person.[115] Some biblical scholars tied Adam to Enkidu.[116] Like Enkidu, Adam lost his innocence when he ate of the forbidden fruit, that is to say the pomegranate. Both Enkidu and Adam were unwittingly transformed from innocence to responsibility and hardship. The agent of Enkidu's transformation was a prostitute and of Adam's reversal a serpent through Eve's enticement. The lasting result for humankind was sexuality, pain, and clothes.

DATE PALM AS *HULUPPU* TREE

In part, my interpretation does not rest on whether the fruit was a pomegranate, although pomegranate is the better case, bordering on certainty. Other trees are suggested as possible identifications of the enigmatic *huluppu*. One other candidate, date palm, has qualities similar to the pomegranate. The case can be made for the *huluppu* tree as a date palm. The inscription of Inanna making her bed and throne from the tree's wood is more compatible with the palm than with the pomegranate, which is a small tree or large shrub. The iconography of early pictures supports both the date palm and the pomegranate as probable identifications. In reliefs, jewelry, and other artifacts of the ancient world, the pomegranate often was associated with the palm.[117] In addition to those representations of Inanna with the date palm presented earlier, there are more examples. Among them is a stele of Ur-Nammu, from Ur, dating from the Third Dynasty, ca. 2050–1950 BCE. The King Ur-Nammu (identified by inscription) pours a libation over a date palm before a goddess on the left. The two deities (shown in the picture) wear the multiple-horned miter and distinctive robe and sit on a throne in the form of a temple. One goddess gestures toward the king and gives him a rod and ring, symbol of earthly power, perhaps corresponding to the *pikku* and *mikku* in the story. Symbolically the king is the gardener to the goddesses.[118]

The date palm has contraceptive qualities. Interestingly, the date palm was one of the first plants that caused modern science to realize that plants produce compounds that stimulate or replicate sex hormones in mammals. In 1933 Adolf Butenandt and H. Jacobi first reported plant estrogens. The plant they reported as possessing "female hormones" was the date palm, a scientific fact since verified.[119] In 1939 Butenandt would receive a Nobel Prize in chemistry, but not for the small article he wrote in 1933 on the date palm. The date palm, he and Jacobi reported, had a compound that they called α-follicle hormone similar in action to human female hormones. What they were describing was an estrogenic compound. Estrogen, of course, is a hormone that promotes or interferes with fertility depending upon timing and amounts.

One problem about the date palm as the tree indispensable to Inanna is that very little use of the date palm as a fertility-limiting agent can be found in antiquity. The Ebers Papyrus (*ca.* 1550–1500 BCE) has a prescription using date palm that is for "loosening a child in the belly of a woman," which can be translated more likely as an abortifacient than as a contraceptive (although both are possible interpretations).[120]

The Babylonian Talmudic treatise *Sabbath* has a probable mention of date palm as an antifertility plant. The discussion concerns drugs that are forbidden on the Sabbath. A "man," however, may eat any kind of food "as a remedy" or drug except water of palm trees and a cup of roots "because they are [a remedy] for jaundice." The cup of roots is clearly abortifacient as is mentioned a number of times.[121] The rabbis explain that water of palm trees is so called because it is spring water that issues from between two palm trees, with one rabbi saying from two different kinds of palms. The rabbis explain that palm water is drunk because it allows the gall to function but it must be taken for forty days.[122] The forty-day treatment would explain why it was allowed to be drunk on the Sabbath. The "water of palms" is unclear unless one postulates that it was a water extract from the palm. If the rabbinical explanation was incorrect in identifying the substance as merely water coming from between two palms and if, instead, the water of palms is a water extract from the palm tree, the Mishnah was speaking of a contraceptive plant. Reasonably, the rabbis were unaware of the pharmaceutical preparation when they loosely defined water of palms.

Still the Talmudic passage and the single recipe in the Ebers Papyrus are meager indicators for what, if the thesis is correct about the *huluppu* tree, must have been a widespread practice. Let us propose that the tree may first have been a date palm (as is indicated by some physical evidence), but that the pomegranate's superior contraceptive effects gradually replaced the palm. Still the earliest representation from the second quarter of the fourth millennium, as mentioned before, was the Uruk vase as votive offering to Inanna with pomegranate clearly depicted. In time, the representation of the fruit associated with the "love" or "fertility" goddess in ancient West Asia may have changed from the date to the pomegranate.

Even Kramer's proposal that the *huluppu* tree might be a willow should be rejected. R. Campbell Thompson, a noted scholar Assyrian of botany whose pioneering efforts are no longer considered definitive, identified *esi (*variant, *ú-su-ú, esu*) as meaning willow; also the plant is called the *hand of Ištar*, Ištar being the Assyrian equivalent of Inanna.[123] And it was used in a pessary for a female disorder, likely for stimulation of menstruation,[124] a common and ancient circumlocution for abortifacient.

On the other hand, *huluppu* occurs in a list of vessels containing drugs and is associated in another Assyrian text as a drug for sorcery[125] and in yet another text as being in the same class with tamarisk, laurel, pomegranate, fir, willow (*şarbatu*), and palm.[126] One Assyrian text has *huluppu* in a beer drunk for childbirth.[127] Dioscorides and other classical ancient medical writers say that the willow was a contraceptive, and modern science has verified the ancients' assertion.[128]

Laurel was known as an abortifacient.[129] Thompson's identification of the *huluppu* in the Assyrian texts, however, must be understood as tenuous at best, because of the difficulty in ascribing plant nomenclature across the time periods of Old Babylonian or Sumerian, Assyrian, and modern. Evidence clearly suggests that the Assyrian *huluppu* is regarded for its female pharmacy.

Ancillary evidence exists to show that the peoples of the ancient world knew of the accidental abortative effects of ergotism, a toxic condition caused by a fungal microorganism that grows on various plants, most readily on types of rye. From around the seventeenth century of our era, ergot was used as an abortifacient. Ancient peoples knew of its devastating effects. When it contaminated bread supplies, entire communities suffered its effects, essentially the same as the synthetic LSD as the modern acid is derived from this biological fungus. The medieval name for the phenomena was St. Anthony's Fire. It is likely that premodern populations did not deliberately employ ergot except in specific rituals and fertility suppression.[130]

Ergotism strikes during certain seasons and requires a cold winter. Robert Biggs found passages in cuneiform texts that indicate that the ancient peoples, by keeping records of weather conditions, foresaw plagues or infestation periods when problems would come. A chief characteristic of these periods was miscarriages. A seventh-century (BCE) text foretold: "There will be recovery for sick people in the land; pregnant women will carry their fetuses full term."[131] And, more ominous: "There will be an epidemic of *li'bu*-disease in the country; pregnant women will not carry their fetuses full term."[132] In Greek literature there are many allusions to a disease of sterility.[133]

TEMPLE SERVICE

Just the same, the evidence as we know it gives meaning to the idea of the young woman seeking protection and control as she enters maturity, the period of mating and childbearing. Herodotus (see below) and numerous other ancient sources, some biblical, describe in various detail how the "love temples" were places where males went to worship through sexual liaison with a priestess or even a priest. The sexual activities transpired in fertility goddess' temples throughout the ancient ages those of Sumerians, the Akkadians, the Babylonians, and other ancient peoples; occasionally even the Hebrews alluded to them in as many biblical passages, especially the Book of Jeremiah II.[134]

OVERPOPULATION

The Atra-Hasīs Epic is an Old Babylonian story about the gods, creation, and early humankind; the text is found in a version written approximately 1600 BCE that tells of how the gods saved humans from a series of draughts, starvation, soil exhaustion, and disease induced by human overpopulation. After consulting with his fellow gods, Enlil announces: "But the people are not diminished./They have become more numerous than before."[135] The gods' purpose in sending the flood was to rescue humans from overpopulation. This second creation, later retold in Genesis as the Noah story, was the gods' benevolent means of saving humanity from too many people to compete for resources. The Sumerian Flood story relates the distribution of duties to the gods Sin and Nergal: "[You should] guard the middle earth./[But you let] loose abundance for the peoples."[136] A. D. Kilmer's study of the Atra-Hasīs text within the context of other myths concludes:

> That limiting the population by "birth-control" should have been conceived by the mythographer as the original motivation for certain cult practices is unexpected, to say the least. The practices to which I refer, and which are known from many contexts, are chastity or celibacy, or methods of intercourse that would avoid pregnancy.[137]

Of course, the avoidance of pregnancy is contraception. In light of the evidence, we should no longer refer to the Mesopotamian institutions as "love temples" or "fertility cults," but we should speak of them for what they were: fertility-control temples and cults.

The instructions that women received in the fertility-control temples and the birthing temples probably were practical and had to do with information that they would need for their marriages. Namely, they learned how to be happy and to make their husbands the same, how to plan the family, and, presumably, how to care for the children they wanted. The women in the temple knew secrets they taught to the initiates.[138] The secrets were those learned by Inanna in her sojourn in the Underworld and by those women who followed her in a struggle against the serpent who could not be charmed, the dark maid Lilith, and the Anzû bird. Since ancient women learned about the *huluppu* tree, women approached maturity and marriage with joy and fear, the latter brought on by the threat of childbirth. Ancient women learned the secrets from priestesses who knew the ways to control reproduction. Modern women have similar controls. Just as with Inanna and her ancient followers, modern women have forces in their societies that seek to remove their use of birth control knowledge. Eve and Inanna are closely connected to modern women.

Herodotus and Sacred Temple Prostitution in Inanna's Court

Herodotus wrote about his travel to Assyria about 440 BCE or, at the very least, he says that he went there, and modern scholars do not disbelieve him on the extent of his travels. In the context of his Assyrian travels (he refers to the people also as Babylonian) he said that he would not tell everything that he saw, predicting that most of his audience (and readers?) would regard many of the stories as made-up.[139] One wonders then why he included the description of young Assyrian women who had to endure intercourse with the first strange man who asked. Herodotus describes the scene:

> The foulest Babylonian custom is that which compels every woman of the land once in her life to sit in the temple of Aphrodite [Greek for Sumerian Inanna and Assyrian Isthar] and have intercourse with some stranger... Most sit down in the sacred plot of Aphrodite, with crowns of cord on their heads; there is a great multitude of women coming and going; passages marked by line run every way through the crowd, by which the stranger men pass and make their choice. When a woman has once taken her place there she goes not away to her home before some stranger has cast money into her lap and had intercourse with her outside the temple; but while he casts the money, he must say, "I demand thee in the name of Mylitta [Μύλιττα[140]] (that is the Assyrian name for Aphrodite)."[141]

After explaining that the amount of money was immaterial to the transaction, he explains,

> So she follows the first man who casts it and rejects none. After their intercourse she has made herself holy in the goddess's sight and goes away to her home; and thereafter there is no bribe however great that will get her. So then the women that are fair and tall are soon free to depart, but the uncomely have long to wait because they cannot fulfill the law; for some of them remain for three years, or four.[142]

Herodotus' Greek audience (we are presuming that his text was publicly read) must have laughed at that passage.

Some scholars in recent years challenged the interpretation of temple prostitution despite Herodotus's account to which is added testimonies of numerous ancient sources, predominantly among them are Strabo, Eusebius, and above all many notices in the Hebrew prophets.[143] Julia Assante argues that the Mesopotamians did not regard prostitution, temple or secular, "as a profession because sex was not considered to be a learned or specialized skill." Some words translated as prostitute, Assante believes, were adult females who did not live with their fathers, some of whom were chaste and others designated as knowing the penis.[144] I find Assante's position unlikely because the weight of the textual evidence sustains the early existence of females whose livelihoods were sexual favors, which is the definition of a prostitute.[145] Stephanie L. Buden presents an even more recent and thorough study than Assante. Her well-documented book's thesis is expressed in the first sentence: "Sacred prostitution never existed in the ancient near east or Mediterranean."[146] About Herodotus's Babylon account, Budin compares similarities between Babylon's alleged customs with the Greek Thesmophoria (to be discussed in chapter two). Budin's documentation of well-reviewed passages produces a scholarly tome with so many qualifying nuances than, when one has digested her arguments, one is left with a mixed, highly qualified definition of what is "sacred" and what is "prostitution." A full review of Budin's work by Vinciane Pirenne-Delforge, herself a researcher on the subject, places Budin in the Herodotus is in the "Liar's school."[147] Herodotus was, she believes, making an exaggerated comparison "between Greek sacred and sexual moeurs" with its largely imagined opposite practices of the effete easterners.[148] Regarding Budin's revisionist position, her reviewer says, "Budin's analysis does not end the debate."[149]

My interpretation pomegranate thesis gives more veracity to Herodotus and the numerous other ancient accounts. By the second millennium, long after the Uruk vase and Inanna legend, prostitution existed; some of it was connected to temples whether within the temple confines (as most texts suggest) or contractual with the sexual workers paying to the temple a tithe (as Assante believes). Assante would discount Herodotus's statement that money was incidental to the woman's intercourse.

How It Happened

To the ancient Greeks Herodotus's account was probably titillating wonder. To us moderns, all of this seems bizarre, even perverted. In private letters, bequests, and other written tablets in cuneiform (e.g., Sumerian, Akkadian, and Assyrian) on clay, we see a family-oriented society that valued clear bloodlines for the transmission of property through wills.[150] A woman, who visited a paramour, was put to death, as well as the paramour, and a woman who had sexual relations with a man in a brothel was subject to severe punishment by her husband. These measures, immoderate to us, were Mesopotamia's way of protecting the family.[151] This concept is in strong opposition to families' requiring their premarriage daughters, presumed virginal, to have semipublic intercourse with the first stranger who requests her in the name of the Good Fertility Goddess.

A fragmented tablet, dated 1727 BCE, relates an extraordinary and seemingly contrary view of virginity: in 1737 BCE a man and woman contracted marriage, but the groom-to-be renounced the obligation. If he could prove his betrothed to be sexually active, he was free from a hefty fine if he jilted his bride-to-be. Alderwomen examined the woman and found her innocent of grounds for her fiancé's charges, but the nature of the charges was unspoken. If we interpret the meaning beyond mere slander, we must assume that he accused her of sexual activity that the committee of women refuted.[152] Assuming the modern interpretation is correct, a woman was expected to be virginal at the time of the wedding. If we view correctly this tablet and initial sexual contact in temple service, we may conclude that the first intercourse in Inanna's service did not "count." The criteria with which the alderwomen examined the women are not explicit.

We know that temples functioned for a variety of purposes (e.g., storehouse for foods, banking, social meals) in addition to giving the gods their due recognition through rituals to be performed. They were also health clinics. For example, in Egypt a woman nearing birth would check in to the temple. Similarly, in ancient Mesopotamia, temples provided health care as part of their function. A deity might be associated with a distinctive specialization.[153] In Mesopotamia, the greatest healing deity was Gula ("great"), the goddess of healing; she was the patroness of physicians. The center temple was at Isin, where she was known also as Ninisina ("Lady of Isin"); her temple was associated with obstetrics and care of neonatal infants (but not to the exclusion of general illness and injuries).[154] Royal inscriptions and tablets attest to the goddess's role in caring for people, notably women, for example:

> She shall be the lady of joy and prayer hastening to the cry (of the people)...
> She shall loosen the band of disease, the distress of tribulation,
> She shall be the counselor, the seer, the magician, the purifier of all things.[155]

And another hymn, having many copies (which make the date of its composition beyond our means), has the goddess speaking in the first person:

> I am a physician, I can heal,
> I carry around all (healing) herbs, I drive away disease.
> I gird myself with the leather bag containing health-giving incantations,
> I carry around texts which bring recovery,
> I give cures to mankind.[156]

Did temple midwives administer parturition at home, or did the near-term woman simply check into the temple? The home was where most therapies were administered, but, at least for the Sumerian period, women nearing birth checked in to the temple where they received obstetrical care from the priestesses who served as midwives.[157] A broken Old Babylonian text reads in translation:

> In the house of child-bearing woman in confinement,
> let brick be laid [a common obstetrical apparatus], seven days.

> Let Dingirmah be honoured, the *sage-femme* Mani,
> Let the midwife rejoice in the house of she in confinement.
> Whenever a child-bearing woman will give birth,
> Let the mother make . . . a baby by herself.
> A male.[158]

One tablet surviving from a king suggests that a pregnant woman (possibly but not clearly the queen) was brought to the temple of Gula. The broken text does not clarify whether the birth took place there.[159] The archaeological remains of healing temples do not provide enough clear evidence as to whether the rooms were for nursing or other temple functions, including examinations.[160]

Generally Sumerologists believe that in third millennium BCE the temple's administrative staffs organized and regulated most work (chiefly agriculture-related), and women were an important part of the work force.[161] Impeding our understanding of everyday women's role is the dearth of scenes of daily life on the reliefs and figurines and other artifacts from early Mesopotamia.[162] In all of the Neo-Babylonian tablets translated, there are approximately 50,000 names of people, but roughly only 2 percent are women.[163] Mostly these would be royalty or slaves. In short we know little about Mesopotamian women. It is mostly through private tablets that we gain information about this subject. As expressed by L. Delaporte, a professor in Paris of ancient Mesopotamia, "for every event of public or private life their [priests and priestesses] good offices were invoked."[164] Myths are generally (always?) gradual and creative deteriorations of what actually happened.[165] Let me propose a means of making sense out of these stories and artifacts.

Conclusion

The earliest foundation inscriptions found at temple buildings, according to Leonard King, are of a "practical rather than of a purely religious character."[166] Just as the Sumerian society provided schools for boys who learned reading, writing, arithmetic, and other skills to enable them to function in a complex business society,[167] let us conjecture that women also were provided some "finishing school," to borrow from a much later American phrase. From the early Assyrian period, we know that women lived in and around temples; some were called *naditu*, literally a "fallow" woman or the "chaste one."[168] *Naditu* women were associated with a temple. Some have interpreted the term to mean prostitute or one thrown down and surrendered to a god,[169] but they appear to be chaste or, at least, childless. In ways beyond our capacity to understand either in the Sumerian, Akkadian, or later Assyrian societies, we learn with incredulity that some *naditu* married but had to remain childless, while others were members of a sisterhood as an alternative to marriage, similar to nuns or, better, *beguines* of medieval Europe. Comparative anthropology reveals that in south Indian Devadasi culture, there were similar temple prostitutes ("slaves of the temple"): high-born women married the temple, were encouraged to have sexual intercourse with "worshippers," but had to remain childless. To be sure, these comparisons are with different cultures and time-periods, but the practices shed some light on a dark subject.[170]

Beatrice Brooks studied cult functionaries and remarked that, if all the Akkadian words that have been translated as "prostitute," "harlot," "whore," and "hierodule" were to be believed, an improbable percentage of the population was engaged as "either secular or religious prostitutes of some sort."[171] That said, an understanding of the historical problem is necessary to understand the relationship between temple prostitution, contraception, and the pomegranate. A similar word, *nu-gig* means "temple prostitute,"[172] whereas *kar-kid* (alt. *kharimutu*) meant simply prostitute.[173] According to Merlin Stone, the Mesopotamian *nadītu* essentially ran the business of the temple, from accounting by scribal writing to numerous business and practical details of running the temple. Earliest examples of writing (*ca.* 3200 BCE) were tablets written by women (*nadītu*) in Inanna's temple, where payments were registered for land rentals.[174] The archaeological findings of Sumerian temples and nobility and common houses, unfortunately shed little light on the organization of the temples, except that the earliest ones seemed complex, self-contained, walled units, but how people related to them is not revealed in the dust and bricks.[175] Buttressing Herodotus's account is an association and word for Inanna in Sumerian and Akkadian. Inanna was called the "Divine Lady Owl" from Sumerian ^{d}*nin-ninna*.[176] The Akkadian word for owl is *eššebu,* and a number of lexical texts give the synonym for "Divine Lady Owl" as *Kilili*, a name also for Inanna (or Ishtar [*Ištar*], Akkadian for Inanna). A prostitute goes out at night like an owl. The Sumerian counterpart of *Kilili* is *Abashushu* that means "harlot."[177] Therefore in the early Sumerian period Inanna was associated with prostitution.

Let us conjecture: under the temple's auspices, women would be taught the essentials of keeping a household, such as household management, pregnancies, birthing, and raising children, in ways similar to the informal functions of women's convents in the Middle Ages. Included in such instruction was how to prevent an unwanted pregnancy, which may have included, if Herodotus is believed, a sacred trial intercourse under the good goddess's auspices and her blessing. In this way, a woman was then ready to be wed. Whether Herodotus and other sources were correct or not, the period of temple service would have been schooling for life's marital and even extramarital experiences. Never would she, her parents, or wooers fear pregnancy by a ritualized but real act. I can think of no alternative interpretation except this one: women learned in the temple how to take pomegranate to prevent pregnancy. To a modern layperson, the knowledge seems simple: take some pomegranate. But how much, what parts, frequency, administered how, when, in relation to the menstrual cycle? All these questions require an empirically derived knowledge of what, in our terms, is hormonal regulation. Estrogen, whose production pomegranates stimulate, can have the effect of enhancing fertility or denying it, depending upon dosage and timing of administration according to the menstrual cycle. Does it make sense for a society that values family, loves daughters and sons, and wants clear blood lines for transmission of name, reputation, and property to the next generation to allow, nay require, its daughters to have sexual commerce with the first stranger who asks even though the action takes places in the temple's precinct? No wonder that Herodotus, a Greek, would not understand but he recognized a tantalizing tale when he saw one. The tale about how Inanna achieved her feminine powers through the use of the *huluppu*'s tree fruit, the pomegranate, preserved the story about how women came to control their reproduction.

Jean Bottéro gives another interpretation of the Herodotus account; he believes Herodotus erroneously stated that all women had to do temple prostitution duty. He regarded ancient Mesopotamian society as one of "free love," by abundance of prostitutes (male and female, incidentally), thereby still protecting the family unit from males seeking gratification. Herodotus, he said, "mistakenly thought they included all the women of the country."[178] Bottéro believes that "it is probable that certain prostitutes, if not all of them, often went to sanctuaries, especially those of their protectress *Ištar* [Inanna] . . . [and exercised] their profession there."[179] Bottéro chooses to ignore (or disbelieve) a critical part of Herodotus's account: "and thereafter there is no bribe however great that will get her" after she returns home from the temple. In other words, the intercourse event was a once in a life-time occurrence. Assante's explanation is better: there was no temple prostitution as such. Most scholars today believe that prostitutes gave a part of earnings to the temple but there were not fertility rituals by the historic period.[180] This is to say that the fertility rituals were not what they were in the Uruk, pre-Sumerian, or even early Sumerian period. Even if Bottéro is correct, especially regarding prostitution in Babylonia when Herodotus visited, contraceptive knowledge would be a virtual necessity for these prostitutes (large in number, he concedes) to serve all sexually active males and to remain pregnancy free. I doubt his theory but, even were it so, my hypothesis remains valid: the temple of Inanna is where women learned to protect themselves through effective contraceptives, and the most effective contraceptive of their time was the pomegranate. Moreover, we learn that usually *nadītu* women were forbidden to have children, and others had to remain chaste in ways we cannot begin to understand.[181] The mother of Sargon, the King of Akkad, was a priestess (*entu*) and forbidden to bear children, as Sargon said, "My father I do not know. . . . My city is Azupirānu [literally 'Saffron'], which lies on the bank of the Euphrates. My mother, a high priestess (*entu*) conceived me, in secret she bore me."[182] Perhaps it is coincidental, but probably not, that saffron was a known contraceptive plant in classical medicine.[183]

One is reminded of the comment made by the Byzantine historian, Procopius (sixth century CE), who described a well-known prostitute: "Naturally she was frequently pregnant, but by using pretty well all the tricks of the trade she was able to induce immediate abortion."[184] Procopius probably understood specifically "the tricks of the trade" and why prostitutes did not bear children at about the same level as in Herodotus' findings. Unless there is compelling evidence, the historian should accept his written sources. In this case, Herodotus probably recorded a combined account of what he saw and was told. Herodotus did not understand how to bring together these fragments of knowledge, other than ancient that Mesopotamian women were instructed in Inanna's temple about the power of the pomegranate, the fruit of the *huluppu*.

CHRONOLOGY

Stages of Civilization	Period	Dates (all BCE)
Hunters and gatherers	Late Paleolithic	15000
Beginning of agriculture and	Epi-paleolithic	12000–7000
early farming villages	Neolithic	8000–4000
	Ubaid	5000–3800
	Early Uruk	3800–3500
Bronze Age begins	Middle Uruk	3500–3200
Early cities in Mesopotamia	Late Uruk and approximate	3200–3100
Development of writing	Date of Uruk Vase	3200–3100
Sumerians and first city-states	Early Dynastic I	2900–2700
(e.g., Kish, Nippur, Ur, Lagash)	Early Dynastic II	2700–2500
	Early Dynastic III	2500–2350
	Aššur (small city state)	2500
Akkad and Akkadians	Early fragments of Gilgamesh and	2334–2160
	Inanna Legends on tablets (legends pre-date tablets)	2150–2000
	Sargon I	2334–2279
Ur III period and Sumerian revival		2119–2004
First Babylonian Dynasty		1894–1595
	Hammurabi	1792–1750
Assyria	Independent state	1300
	Assyrian power	1000–612

Source: Adapted from tables in Robert Chadwick, *First Civilizations: Ancient Mesopotamia and Ancient Egypt*. 2nd ed. (London: Equinox, 2005).

CHAPTER 2

POMEGRANATE AS EVE'S APPLE

> [*Genesis* 2: 8] And the Lord God planted a garden in Eden, in the east; and there he put the man whom he had formed. [9] And out the ground the Lord God made to grow every tree that is pleasant to the sight and good for good, the tree of life also in the midst of the garden, and tree of knowledge of good and evil. [18] . . . Then the Lord God said, "It is not good that the man should be alone; I will make him a helper fit for him." [God makes birds and animals, "every living creature" and has Adam to give them names.] [21] . . . So the Lord God caused a deep sleep to fall upon the man, and while he slept took one of his ribs and closed up its place with flesh; and the rib which the Lord God had taken from the man he made into a woman and brought her to the man. [25] . . . And the man and his wife were both naked, and were not ashamed. [3:1] Now the serpent was more subtle than any other wild creature that the Lord God had made. He said to the woman, "Did God say, 'You shall not eat of any tree of the garden?' " [2] And the woman said to the serpent, "We may eat of the fruit of the trees of the garden"; [3] but God said, "You shall not eat of the fruit of the tree which is in the midst of the garden, neither shall you touch it, lest you die." [5, serpent to Eve] "For God knows that when you eat of it your eyes will be opened, and you will be like God, knowing good and evil." [6] So when the woman saw that the tree was delight to the eyes, and that the tree was to be desired to make one wise, she took of its fruit and ate; and she also gave some to her husband, and he ate. [7] Then the eyes of both were opened, and they knew that they were naked; and they sewed fig leaves together and made themselves aprons. [Revised Standard Version cited unless otherwise noted.]

Already raised is this question: is there a connection between Inanna (who was in the Underworld with the god of knowledge, and her possession of the *huluppu* tree) and Eve, the serpent, and eating of the fruit of the tree of knowledge? In a word, the answer is "yes," and it will take much of this chapter to explain why. As I set out in chapter one to explain, Inanna and her association with pomegranates dates back at least to late Uruk (ca. 3200 BCE), and that Inanna and the *huluppu* (pomegranate) probably existed in legend to at least that time. If the tablets are later than the vase, you ask, how can we know that the legend as told in the tablets dates back any earlier than the tablets? The way we know is found in the art of philology. If, for example, we had only fragments of texts from the twentieth century and, among them were sections from Grimm's fairy tales, a scholar a thousand years later, say in the fortieth century, could tell that the language and subject matter (as examples: princes, dragons, castles, fairies, horses for transportation) were from an earlier era, one we call the Middle Ages.

Composition of Garden of Eden and Fall Stories

The Book of Genesis in the Torah (or Bible) is a conglomerate of various texts, written by different people over periods of time spanning centuries; the writings as we have them are based on traditions, myths, and histories in human memory that predate the actual written composition. Preliterate societies are fairly good at transmitting knowledge, better often than literate societies whose minds are trained in information storage and retrieval, not memory. Genesis 2: 4b-25, including the passage above, covers the creation of man and woman, the Garden of Eden, and Fall, and the Tower of Babel. Scholars refer to the section as "J," named from *Jahwist* (or *Jehovist* from Hebrew *Yahweh* = YHWH); throughout the section this name represents a human-like God. Its composition (roughly in the form that we have as transmitted) dates from as early as Israel's early monarch (*ca.* 950 BCE and later incorporated into the Torah (*ca.* 400 BC). A number of scholars see evidence of composition in the pre-exile, exile, or early restoration period, known as the Babylonian Captivity (*ca.* 600 BCE- 516 BCE= rebuilding of the temple).[1] Certainly—and, yes, that is the correct adverb based on sufficient evidence—the early sections of Genesis dealing with Creation (1:1–2:4a), probably composed in the post-Exile period, and "J" source, were written after Jewish familiarity with Mesopotamian cultures. Before the Exile, the Jews in Palestine would have heard of Asssyrian expansion to their northeast. Through the victory of the Neo-Babylonian King Nebuchadezzar (r, *ca.* 605–561 BCE), Judea became part of or subjected to Babylonian rule; in 597, the Judean King Jehoiachin was forcefully deported to Mesopotomia alone with his court and retinue, numbering in the thousands, thus beginning the so-called Babylonian Captivity. After a subsequent rebellion Nebuchadrezzar destroyed the temple in Jerusalem (586 BCE) and exported more Jews. After forty-eight years in Mesopotamia, the Persians under Cyrus the Great conquered Babylonia and allowed the Jews to return. An important detail is that not all Jews were exiled, many remaining in Palestine, and, once those in Mesopotamia were freed to return to Palestine, some remained in their new homes between the Tigris and the Euphrates Rivers. Just the same, many Jews rejected Babylonian culture, religion (e.g., polytheism), and societal conventions (e.g., ritual and temple prostitution, as in *Jeremiah*), but, importantly, they also absorbed some of its conventions and legends; some Jews greeted Cyrus as being "anointed" by God (*Isaiah* 45:1).

In imagery, language, and context the creation story has parallels with the Sumerian Epic of Creation (*Enuma elish*; *Atra-Hasīs*), the "ladder" (Gen. 28:12) with the Mesopotamian *Nergal and Ereshkigal* myth, the "Sons of God" with Ugaritic "Sons of God" or "*Sons of El,*" and the Noah Flood story with Ziusudra, the Sumerian hero of the Flood (also called Utanapishtim in the *Epic of Gilgamesh*).[2]

An example of association (or borrowing) comes from the Sumerian myth of *Adapa and the Flood of Life*, where: "He [Ea, the Father God] granted him [Adapa, his son] wisdom, but he did not grant him eternal life"[3]—the same punishment that Adam and Eve receive for disobedience. Without exhausting the list of parallels between the Torah's content and Mesopotamian culture, and before we examine how the pomegranate relates to and, indeed, is the "fruit of the Tree of Knowledge," permit me a few words about reading the "Bible." Scholars use the Greek word *exegesis* ("to explain"; "to interpret") so that they can try to understand the author's meaning within the context of his culture and *eisgesis* ("to interpret" in light of one's own meaning and time).

Analogy is the weakest form of argumentation; I begin with analogy. Lynn White, professor of medieval history at the University of California, Los Angeles, was at home in the early 1960s—he told me, thus only orally transmitted until now—when a telephone call came from the producer of the movie version of "Camelot." Prof. White was asked to be the historical advisor for the movie version of the Broadway play. He gave a short introductory lecture on the historiography of the Arthurian legend: if there had been a King Arthur, he would have been a sixth-century Britano-Roman who defended his region against the invasion of the Angles, Saxons, and Jutes.

The name Arthur may appear in a short passage in Gildas's work (mid-sixth c.) and then not again until Nennius's late tenth-century mention. Notably the name is absent in intervening histories such as the one by the Venerable Bede. Around 1100, Geoffrey of Monmouth wrote a history of British kings, in which he took legends, probably mostly in old ballads no longer extant, and possibly but unlikely, some lost written works and wove together an exciting narrative history, so exciting that the Romance epic writers from the mid-twelfth through the thirteenth century used the stories of the knights and ladies of the Roundtable, the stories that inspired the adaptation of a musical based on T. H. White's novel (1958) the *Once and Future King*. Professor White explained that there were so many layers that it would be impossible to advise historically on such a movie: what costumes would they wear, sixth c., ninth c., twelfth c. or thirteenth century and more to the point, what values would they espouse, the masculine roughness of sixth-century warriors or the gentlemanly behavior of Gawain, Yvain, Lancelot, and Perceval, and ladies fair, such as Genevieve, who were treated by a new knightly code called chivalry? Since those medieval poets, almost every generation has produced literary works (and, more recently, motion pictures) using the Arthurian matrix. From my youth the values have changed with each retelling (viz. Mary Stewart); the most recent novels and movies that I have seen reflect the political correctness of the late twentieth century (such as a respect for diversity). Professor White summarized by saying "go ahead and make the movie because there is no way to make it historically accurate." When the movie appeared, he was credited as its adviser but, in his words, the honorarium was so large that he dared not protest.

In interrupting the Genesis' "J" component, we need to go beyond the *exegesis* but not to exclude an understanding of its meaning to us in the post Roe-versus-Wade era. By going beyond, we gain some insight into its meaning; we study various thinkers and the general culture, from the time of its composition to today, especially the Talmudic opinions, Latin and Greek Church Fathers, and medieval scholastics. The Qur'an (4:1) alludes to the Garden of Eden story, but with no explicit mention of a fruit, tree, or disobedience. The author of the "J" component of Genesis was obviously exposed to Mesopotamian lore, but the Hebrew author reverses some Uruk and Sumerian values. In so many ways the Hebrew religion was asserting a new, anti-Mesopotamian direction. Religion embraces politics, so has it always been. In Genesis, God is a single male and not the principal deity as a fertility goddess, as the deity in Uruk and the cultures that followed. In Genesis, even the birth of woman comes from a man, Adam's rib, and Adam gave Eve her name, thereby demythologizing the Sumerian stories about Inanna's creation.[4] A geographical catalogue of the rivers that form from the Garden of Eden (Gen. 2: 10–14) includes the Tigris and Euphrates. The issue in "J" is the state of innocence in the Garden, God's prohibition of eating the fruit of knowledge without an explicit statement as to why it was forbidden, the temptation, the meaning of the serpent, the enticement of Eve and subsequent enticement by her of Adam, and the expulsion from the Garden, the liability of death, the pain of childbirth, and, to the consternation of many moderns, the subjugation of a wife to her husband: "he shall rule over you" (Gen. 3: 16).

Eve as First Woman

The name Eve for the first woman comes from the Hebrew word *ḥawwâ* because she is "mother of all living (*ḥay*)." Etymologists, however, cannot connect with certainty Eve's derivation with the root *ḥyh,* meaning "to live."[5] In Sumerian cuneiform symbols the word for "life (*til*)" also means "rib (*ti*)," allowing a double meaning of Eve as the Mother of Living and Mother of the Rib.[6] Through Ugaritic and Phoenician, some philologists have tried to connect Genesis' Eve with other female fertility deities of Ancient West Asia, including the Akkadian goddess Mami and, inferentially, the Sumerian Inanna.[7] Mami was also known for her midwifery skills and, according to one study, she shared qualities with Eve and Pandora: each was involved in creation stories that account for the origin of evil.[8] The author of the Garden of Eden story may have associated Eve with a foreign fertility goddess, but even if not, the point is that the Genesis retelling secularizes her as human and of one flesh with Adam.[9]

Genesis 2:9, 17, describes the tree as "the tree of learning," "that which is to be known," or "the knowledge of good and evil." The Hebrew stem *yd`* is not only "to know" but "to experience, to come to know." The delicate phrasing, compared with similar idiomatic expressions in the Bible, implies that Adam and Eve could not distinguish "good and bad," as they knew not the shamefulness of their nakedness, until the forbidden fruit. The learning of physical aspects of life, according to biblical scholar E. I. Speiser, leads to the mystery of sex.[10] Another scholar, J. Coppens, regards Eve's transgression as sexuality tempted by the serpent, a phallic symbol (later to be discussed).[11] If the fruit was the pomegranate and its meaning contraceptive, Genesis "J" 's meaning is even deeper. All the more, Eve is connected with Persephone who discovers the secret to thwart a man's will to force pregnancy.

God in Genesis and Zeus in Greek myth are equally angered by the woman's (before godliness) discovery of a god's secret.

In placing Adam and Eve in the Garden, God's one prohibition was "thou shalt not eat" of the "tree of knowledge of good and evil." The temptation that comes directly from the serpent is later associated with the devil, but the association was made by later writers.[12] The pseudoepigraphic writing, "Life of Adam and Eve" (*ca.* first century CE) and known in Latin and Greek (no Hebrew), has little information about what it calls "the tree of the knowledge of good and evil," but in it the devil is a separate god who uses the serpent as a tool to achieve his objective to cause humans' downfall.[13] The "J" author provides us with no reason that the serpent was the tempter, but delivers it a compliment: "more subtle than any other wild creature that the Lord God had made" (Gen. 3:1). In Palestine there are thousands of fertility figurines, usually a naked woman with a serpent entwined around her neck. Also the serpent is frequently depicted on cup handles and other artifacts made by the ancient Hebrews.[14] To the ancient Sumerians and Egyptians, as well as to the Canaanites—the Jews' likely intermediary source—the serpent was a fertility symbol,[15] but the "J" source for Genesis does not make fertility explicit. Indeed, the serpent's function relates to the structure of the narrative as a means to unfold the action; because of its cultural familiarity, the serpent was convenient, not the central point: the point is humans' disobedience to God that causes the "fall" from grace and the loss of immortality and innocence. In eating the forbidden fruit, humans obtained knowledge of good and evil, an act of *hubris* that implies equality or even rivalry with God. The snake in the roots was the near-universal symbol in the West Asian ancient world of evil, immortality, and sexuality.[16] The snake was associated with fertility themes in ways that are obscure. The Aramaic word for serpent, *ḥewyā*, is related to the Hebrew word *ḥawwâ*, meaning Eve, and it is connected to the mother goddess Asherah Ṭannit.[17] The Hebrew word for serpent has the same root as the word for life.[18]

In the Epic of Gilgamesh, the plant of life was stolen from Gilgamesh by a serpent, foreshadowing the Genesis story.[19] One interpretation is that the serpent symbolized reoccurring youthfulness or immortality, perhaps through its rebirth process in shedding its skin. As a combination of wisdom and chaos, the serpent sought to persuade men to become gods.[20] Another interpretation of the serpent in Genesis and Mesopotamian usage (or, at least, non-Israelite) is that the serpent was a phallic symbol.[21] The Genesis writer, however, did not know Sigmund Freud. John L. McKenzie interprets the serpent's role in Genesis 3:14–19 as "nothing else but that sexual life is a curse to the woman."[22] Of course, Genesis' meaning may not be the same as in the Inanna story, but, whatever its meaning, Inanna was disturbed by the presence in her tree of the serpent "who could not be charmed." In Genesis, the serpent lured Eve into eating the apple, she lured Adam, and they both discovered their nakedness and were expelled from the Garden. St. Ambrose, like most Latin Christian Church Fathers, saw no particular meaning in the fruit except that it imparted reason, understanding, and knowledge, whereas before eating it Adam and Eve were "naked because of the purity of their virtue." Being a misogynist, Ambrose saw particular fault with Eve; her deception cause Adam to fall "by his wife's fault, and not because of his own."[23] The Church Fathers understood the lesson to be one of human innocent purity and then disobedience. The "fruit" could have been anything that God could have used to unfold humans' disobedience and fall.

Tree of Life

The pomegranate is frequently named the "tree of life" on Neoassyrian seals, thus the correspondence between Genesis' "tree of knowledge" [24] Just the same, the Genesis passage makes a clear distinction between the tree of life and the one of knowledge, of good and evil.[25] There are two trees in the Garden, one the tree of life, the other the tree of knowledge and evil (*Gen.* 2: 9). A recent study of the archaeological records that represent the tree of life demonstrates that, far from being mythical, the Egyptian lotus plant was the tree of life but, in time, its contact with the gods grew dimmer.[26] The Genesis theme centers on the tree of knowledge and evil, but the tree of life is mentioned once more after the expulsion from the Garden and the Fall. In *Gen.* 3:22, the narrator refers to the tree of life to say that, after having eaten the forbidden fruit of the knowledge tree, humans would have gone on to sample its fruit and gain immortality: "Then the Lord God said, 'Behold the man has become like one of us, lest he put forth his hand and take also of the tree of life, and eat, and live for ever.' " Whether or not the tree of life was conflated with the *huluppu* tree is not directly pertinent to the Genesis story. The "J" author of Genesis distinguished between the two trees. He employs the tree of life as a supplementary or ancillary point about why humans must suffer and die. The essence of the story is how people came to be as they are.

The immortality theme ran throughout ancient culture just as it did in the much-earlier Gilgamesh Epic (date *circa* 2700 BCE), where the hero sought immortality from the plant of life ("this plant is a plant *apart* from any other") at the bottom of the sea.[27] The supposition is that Gilgamesh's "plant of renown" was the ubiquitous "tree of life."[28] Gilgamesh said, "I will give it to the old men to eat. Its name shall be 'The Old Men are Young Again,' and I shall eat it myself and have all my lost youth." Gaining the plant of life by diving to the sea bottom, he encountered "a serpent... [which] rose out of the water and snatched it away and immediately sloughed its skin and returned to the well."[29] Long before the Greek tragedies, Gilgamesh was guilty of *hubris*, "overweening pride," and he was duly warned by a god, only to have a serpent rob him of it. Siduri, the goddess of the garden of the sun, said to him:

> You will never find the life for which you are looking. When the gods created man they allotted to him death, but life they retained in their own keeping. As for you, Gilgamesh, fill your belly with good things; day and night, night and day, dance and be merry, feast and rejoice. Let your clothes be fresh, bathe yourself in water, cherish the little child that holds your hand, and make your wife happy in your embrace; for this too is the lot of man.[30]

No evidence points to the Genesis "J" author's having direct knowledge of the Gilgamesh Epic as we have it preserved on tablets, but the story line was part of the general culture of Ancient West Asia. As evidence of familiarities and similarities, Siduri's advice is repeated virtually the same way in Psalms 113:117 and Ecclesiastes 5:18, 8:15, and 9:8–9.[31] The tree of life is peripheral to the Fall; it appears to account for man's mortality within the author's purpose to discuss why humans, in disobeying God, were punished and expelled from the Garden. The issue unfolds around the prohibition of eating from the Tree of Knowledge, the Forbidden Fruit. What was its meaning?

FRUIT OF THE TREE OF KNOWLEDGE

The fruit of the Tree of Knowledge was special: the pomegranate. Why were pomegranates prohibited? Some scholars place only symbolic value in the actual fruit because by their interpretation humans disobeyed God—to St. Augustine and other later Church Fathers accounting for "original sin"—so any prohibited rule was as good as another, such as "don't step on cockroaches." The Genesis meaning of "to know" goes with "good and evil," but good and evil what? Biblical scholar, Howard N. Wallace summarizes the scholarship with three interpretations: (1) a claim of the right of self-determination, a form of human maturity, that some extend to assumption of god-like powers; (2) knowledge of sexual relations; (3) access to universal knowledge.[32] Most scholarship holds the second position, sexual relations, simply because the Genesis author before and immediately after the introduction of the prohibition of eating the fruit of the tree placed first the innocent nakedness of Adam and Eve and thereafter the shame for the same. Other scholars believe that the fruit imparted magic, a rival power of God, but such an interpretation is neither explicit nor implicit. Adam and Eve were naked and innocent, a point of emphasis, and, after eating the fruit, they were ashamed and clothed themselves. The knowledge was not physics or mathematics (although in a way, it was biology). Elaine Pagels recently went right to the point: "The tree of knowledge conveyed *carnal* knowledge."[33] Biblical scholar John Skinner used these words: "the tree of knowledge, whose fruit excites the sexual appetite and destroys child-life innocence."[34] Immediately after eating the forbidden fruit [Genesis 3:7]: "Then the eyes of both were opened, and they knew that they were naked." Adam and Eve were not tempted to know the laws of nature, thereby rivaling God, but to know one specific area: sexuality and reproduction. J. M. Evans, a scholar of the ancient world, wrote: "The use of the phrase 'to know good and evil' elsewhere in the Old Testament[35] suggests that the kind of knowledge prohibited to them was scientific rather than ethical in character."[36] Evans is correct; the knowledge was scientific and pertained to reproductive physiology.

Did the author of this Genesis section know that the fruit of the tree of knowledge was the pomegranate and, if so, did the author know that the sin in question was contraception and reproductive power? The next three successive sections will show that the fruit was the pomegranate, that the pomegranate became a symbol for the early Jewish people (thus was popular and well-known), and, finally through Greek mythology, we can postulate better the meaning to its author of the story in Genesis.

Pomegranate as Fruit of the Tree of Knowledge

In folklore, art, and literature, the fruit is usually named, described, and drawn as being the apple. An apple it was not, even though the Middle English translation of the apocryphal *Vita Adae* ("Life of Adam") identifies the fruit as an apple: "He forbade Adam an appel-tre."[37] In Hebrew the apple is *tappûᵃḥ* (Ugaritic, *tpḥ*) that means a "gold-colored fruit" of some kind,[38] but Genesis (2:16–17; 3:2) uses a term for generic fruit.[39] The Greek Septuagint translated the term into καρπός (or Roman, *karpos)*, generic for fruit.[40] Although apple (*Malus domestica* Borkh./ *M. pumila* Mill.+ sp.) is often the translation, the apple does not grow in Palestine, Turkey being the southern-most limit in the modern period.[41] Because also there are other terms designating it in biblical works, the common apple is unlikely the fruit named in Genesis.[42] Curiously, in all the ancient art of West Asia, the apple was not to be depicted.[43] Chapter one notes that the Sumerian word *hašhur* and Akkadian *hašhuru*, translated as "apple," was not the common domesticated apple that we today eat to keep the doctor away.

Various other candidates that fit the description of "gold colored" have been suggested: citron (*Citrus medica* L.), quince (*Cydonia ablonga* Mill.), and apricot (*Prunus armeniaca* L.).[44] The "fruit" well describes the apricot, but it is unlikely to have been known in biblical times, possibly not being introduced until Roman times.[45] Overlooked is the candidacy for the pomegranate as Eve's apple.[46] Its fruit is yellowish-brown and its name in Latin means "grained apple" (*pomum*= apple or generic fruit; *granum*= grain, seed). The common English phrase "comparing apples to oranges" is uncommonly wrong, because for long periods of history "apples" were any generic fleshy fruit; such as the orange (*Citrus sinensis* + spp.) was called *pomum de orange* and the lemon was *pomum medicum* ("medical apple"). Pomegranate is an English corruption of "grained apple" from *pomum granatum* and, interestingly, from this base-word comes "hand grenade" due to its similar appearance to the fruit.

Pomegranates as Symbol of Judaism

The pomegranate was an important symbol in early Judaic religion. Aaron's high priest's robe's skirt, "all of blue," was bordered with "pomegranates of blue and purple and scarlet stuff, around its skirts, with bells of gold between them, a golden bell and a pomegranate."[47] The later Jewish scholar, Philo (*ca.* 20 BCE–50 CE), described the high priest's robe in his time as having a pomegranate fringe.[48] In the later *Song of Solomon* (4: 3, 13; 6:7), the pomegranate flower was a symbol of spring. The calyx of the fruit was the motif for the crowns of the Torah in decorations.[49] Like no other fruit, the pomegranate was central to Judaic imagery. Josephus (*Antiquities of the Jews*. 3. 7. 4) repeated the description, but, in *Wars of the Jews* (5. 5. 7), Josephus (writing c. 75–94 CE) added an interpreted detail: the bells signified thunder and the pomegranates lightning.[50] Likely, Josephus's interpretation is based on the play on words, *rimôn*, Hebrew for pomegranate, and *Rimmôn*, a god wielding the thunderbolt.[51] The Israelites and early Christians held to a tradition that when Adam was leaving the Garden, he cut a branch from the Tree of Knowledge to form a rod, the same rod that Moses would later carry.[52]

POMEGRANATES AND THE PROMISED LAND

While wandering in the Wilderness of Zin (probably between the Dead Sea and Gulf of Aqaba), the Israelites said to their Lord:

> [*Numbers* 20:5] And why have you made us come up out of Egypt, to bring us to this evil place? It is no place for grain, or figs, or vines [grape], or pomegranates.

Indeed the Land of Palestine, the "Promised Land," was characterized as where grain, figs, grape, and pomegranates grew:

> [*Deuteronomy* 8:7–8] For the Lord your God is bringing you into a good land, a land of brooks of water, of fountains and springs, flowing forth in valleys and hills,[8] a land of wheat and barley, of vines [grape] and fig trees and pomegranates.

Far more than a refreshing fruit with a sweet juice, pomegranates were special to the Jewish people.[53]

POMEGRANATES AND THE TEMPLE IN JERUSALEM

Before the front door of King Solomon's temple was two free-standing columns, eighteen cubits high, adorned with pomegranates in two rows round the capital as on a bowl, four hundred pomegranates per row with two rows.[54] The description in I Kings is

> [7:15] He [referring to Hiram of Tyre] cast two bronze pillars. Eighteen cubits high [that is, about 27 feet] was the height of one pillar, and a line of twelve cubits measured its circumference; it was hollow, and its thickness was four fingers; the second pillar was the same. [16] He also made two capitals of molten bronze to set on the tops of the pillars; the height of one capital was five cubits; the height of the other capital was five cubits. [17] Then he made two nets of checker work with wreaths of chain work for the capitals upon the tops of the pillars; a net for the one capital, and a net for the other capital. [18] Likewise he made pomegranates; in two rows, round about upon the one network, to cover the capital that was upon the top of the pillar. [21] . . . He set up the pillars at the vestibule of the temple; he set up the pillar on the south and called its name Jachin; and he set up the pillar on the north and called its name Bo'az.

The architecture of two columns standing before a temple can be dated back to as early as approximately 1500 BCE where two pillars stood before the temple at Byblos. Later coins show two columns before temples in various locations in ancient West Asia. The pillars' names, Jachin and Bo'az, lead philologists and other scholars from one theory to the other with no consensus among them for many centuries. The symbolic meaning has even more proposals, ranging from phalli, trees of life, obelisks, and twin mountains. To the confusion is added the fact that the Jachin and Bo'az columns have secret meanings in Freemasonry. Among the theories is the etymological tracing of Jachin to the Sumerian fertility god, Nin-gis-zida, addressed as "Lord of the Tree of Life" and depicted as a serpent with a human head, and Bo'az as a corruption of Tammuz or Dumuzi, Inanna's consort.[55] One prominent theory is that the two columns represented male and female to the ancient Israelites.[56] The importance of the two columns lies at its top, with the pomegranates. At the very entrance to the temple in Jerusalem there were pomegranates. "Those ye' who enter therein"—to paraphrase Dante—were aware, I submit, that the pomegranate was the fruit of the tree of knowledge and made them the responsible, sin-making, sex-loving, pain-bearing people that they were after the disobedience and fall. One certainty that we have, a point on which all scholars can agree, is that religious art of the ancients had meaning and was not a product of random, chaotic lines and drawings, however aesthetically presented. The meaning of this art to those who entered the temple in Solomon's time, however, may not have been exactly the same as to those who entered the temple in, say, the time of Julius Caesar, when the temple still stood on the hill in Jerusalem.

Ivory, Temple in Jerusalem, and Ancient Iconography

An antiquities dealer in Jerusalem acquired—exactly how is unknown—an old hippopotamus ivory shaped as a pomegranate.[57] The pomegranate vase is slightly less than two inches tall and one inch wide, on which was carved an inscription in fifteen old Hebrew letters. In July of 1979, André Lemaire, an expert paleographer, examined the inscription and reconstructed these words in translation: "Belonging to the tem[ple of the lo]rd, [Yahweh], holy to priests."[58] Ultimately the Israel Museum purchased the vase for slightly more than a half million dollars; a controversy about its authenticity ensued. Its provenance is unknown; the antiquity dealer had a reputation for unscrupulous sales. If the pomegranate vase's inscription was as authentic as Lemaire stated, it would be the only tangible object that survives from the temple of Solomon, and probably worth far more than a half million.[59]

Now we know that the vase is authentic but its inscription likely is a forgery. Previously the Museum's director, Nahman Avigad, was uncertain about whether and, if so how, the vase was involved in the temple's priestly procedure and ritual.[60] The pomegranates on the priests' robes and on top of the columns before the temple indicated to him a probable temple use. Recently laboratory examination of the so-called Solomon vase proves its antiquity but far earlier than the alleged-bogus inscription suggested: the pomegranate vase is from the Late Bronze period (thirteenth–twelfth centuries BCE). Scholars of ancient Israel concluded that its inscription was forged in modern times, but laboratory testing proves that the vase was much earlier than previously thought. Whereas most investigators believe the inscription to be a forgery, the opinion is not unanimous. Whoever the forger was, he knew much about early Israeli culture because of the plausibility of the pomegranate's association with the temple. Another pomegranate vase (called the Rimmon Vase, from the Hebrew word for pomegranate) was found in Tell Halif in southern Judea and dates from the tenth to the eighth centuries BCE—later than the so-called Solomon Pomegranate Vase. Also from a Philistine temple within the present city-limits of Tel Aviv a pomegranate shaped terracotta vessel was found that dates to approximately 1000 BCE.[61] Actual dried pomegranates were found in ancient tombs at Jericho in Palestine that dates from the Middle Bronze Age (*ca.* 2300 BCE).[62] The small vase may have been the head of a scepter and is more evidence of the pomegranate's popularity in Palestine cultures, including the Jews, but we cannot directly tie it to Solomon's temple except through representations on the temple itself and priests' robes.

An Israeli seal from either the eighth or seventh century BCE is bordered with pomegranates. Also there were the two pendants in Israeli Megiddo, dating eight century BCE, that have pomegranates shaped like the "House of the Lord" Vase. Their shapes were similar to the chains of pomegranates hanging from an Ugarit cultic tripod dating from the thirteenth century BCE.[63]

JUDAIC COINS AND THE POMEGRANATE

Just as in the modern world, where we associate national symbols with the coins and paper money issued by governments, so too was the practice in the ancient world. For example, Athen's coins often had either Athena or the owl (her symbol); Rome had Romulus and Remus (mythical founders of Rome); and, as shortly we shall see below, Cyrene had the plant silphium. The symbol on Jewish coins was the pomegranate, the fruit that adorned the entrance to the temple and the borders of the temple priests' robes. One of the earliest Jewish coins was issued by Yehonatan (Greek name, Alexander Jannaeus, 103–76 BCE), the great-nephew of Judas Maccabee, the rebel leader who led the Jews to independence from the Greek Seleucid Empire. Yehonatan was the high priest who also assumed the title of king. To distinguish his realm from the neighboring Hellenistic powers, he issued crude coins to show that the Jews had reappeared as a political identity, and he used pomegranates to designate that distinction. Some coins were in Greek, others in Aramaic and Hebrew letters, and with mistakes. The obverse of one typical coin has the Hebrew inscription: "Yehonatan the high priest and the community of Jews" while its reverse has a double cornucopia adorned with ribbons and a pomegranate coming from the middle of two horns.[64] The motif of a pomegranate between horns (as a cornucopia) was the usual symbol of ancient Jewish coins.[65] Later the horns were dropped but the pomegranate remained, as in the Shekel of Israel: Year Two = 67–68 CE. Hebrew legend: "Jerusalem the Holy" Obverse: Three pomegranates on a Shekel coin issued in 67–68 CE, just before the Roman destruction of the temple and the Diaspora. One side has "Jerusalem the Holy" and its reverse three pomegranates.[66] In 2008 the State of Israel issued a two-shekel coin with a pomegranate set in a cornucopia in imitation of the ancient coins.[67]

JUDAISM AND THE POMEGRANATE

From the pomegranates on columns in front of the temple, the priests' robes bordered with pomegranates, numerous pomegranate artifacts found in Palestine, and as the symbol on the coins, the pomegranate was an early symbol for the Jewish people, just as the Menorah was later. Did the early Israelites, like those in ancient Sumer, know that the pomegranate was a contraceptive as well as a good refreshing fruit to eat? Did the sin of humankind in eating the Fruit of the Tree of Knowledge of Good and Evil constitute the assumption of fertility control? Before addressing this question, we need to remember: First, through Greek mythology we have better evidence of the pomegranate as the fruit that enables a woman to control her fertility, and, second, from the medical writings we can show that the knowledge of the contraceptive quality of pomegranate was known to many, that is to say, most, people throughout the Mediterranean region and Ancient West Asia. The pomegranate was a very popular subject for the manufacture of jewels, jars, and various vases throughout the eastern Mediterranean world, including the Mycenaean period of Greek cultures. Many of the pomegranate objects are similar and point to factory manufacturing in Syria and Egypt.[68] The tomb of Amenhotep II (1450–1415 BCE) had no fewer that nineteen votive pomegranates of faïence, and on Cyprus at least sixteen glass pomegranates were found in tombs of wealthy people living in the late Bronze Age; those objects were almost certainly not manufactured on the island.[69]

Persephone and Pomegranate

The Greek counterpart of Inanna is Persephone and her abduction; the story is also associated with pomegranate fruit. A statuette of the archaic period shows a goddess, probably Demeter or Persephone, with a pomegranate in her right hand.[70] In the *Hymn to Demeter,*[71] the early version of the story dates from the seventh century BCE, when the Greek city-states (*poleis*) were in their formative period. Persephone (sometimes *Kore*, meaning "virgin") was the daughter of Zeus and daughter-in-law (later, daughter) of Demeter, the goddess of fertility and earth-mother. Some scholars speculate that, before the written accounts, the mother-daughter pair was one fertility goddess who, for reasons lost in the mist of time, became two.[72] There is a similar hypothesis about Inanna and Ereshkigal, her sister (discussed in chapter one). The Eleusian Mysteries had secret ceremonies at Eleusis, supposedly where grain was first grown and where Persephone returned to the Earth. The secret ceremonies honoring Demeter and Persephone date back to Mycenean times (ca. 1700 BCE). In the beginning of the *Hymn to Demeter,* Persephone appears ready for sexuality, only that of her own choosing, just as Inanna wanted her sexuality in the Underworld.[73] Hades (Pluto to the Romans) seized and kidnapped Persephone to the Underworld. Unable or unwilling to intervene, Zeus told Persephone not to eat while in the Underworld. Demeter was distressed, and earth manifested her distress: all that was green and fertile died. Facing the laments of Demeter and the wilted earth, the other gods beseeched Zeus to rescue her. Before the rescue could be executed, Persephone ate some pomegranate seeds, seven by one account and three by another.[74] (Winters, after all, vary as to length of months.)

Questions arise: Were Persephone and Hades "really" married in the Underworld?[75] Was Persephone raped or merely an unwilling bride or consort? The *Hymn* makes clear that sexual intercourse took place,[76] but under what circumstances? Did Zeus know the meaning of eating pomegranate seeds? Did he know and not want Persephone to eat them? In the *Hymn to Demeter* Zeus sends Hermes to bring Persephone back, but why does he not tell Hermes to instruct Persephone not to eat the seeds? Why does Hermes give her the seed secretly? This line from the Hymn explains, "But he stealthily put in my mouth a food honey-sweet, a pomegranate seed, and compelled me against my will and by force to taste it."[77] The

Hymn is no more helpful in answering these questions than was the Temptation in the Garden of Eden. Was Hermes a tool of Zeus, as the serpent was the same for God?

A number of curious discrepancies about the pomegranate make the story incomplete. First, according to the legend, Persephone ate the pomegranate; the medical writings have it administered as a vaginal suppository. There is only one possible source in which we have found pomegranate as a possible oral-route drug. During the Middle Ages, the Arabic text of Ibn Sīnā (Avicenna, c. 980–1037) prescribed pomegranate as a postcoital contraceptive, whereas the Latin translation made by Gerard of Cremona (d. 1187) implied that it was taken orally.[78]

EARLIEST MEANING OF PERSEPHONE FROM A LATE SOURCE

Persephone's eating of pomegranate concerns women's use of birth control agents more directly than previously known by modern interpreters. Symbolically the story is interpreted today as sexuality and death, and the eating of the man's food as seduction and marriage consummation. Unwillingness, treachery, and rape are implied.[79] A late Roman retelling of the story adds new meaning. Claudianus (*fl. ca.* 400 CE) wrote but did not finish "Rape of Persephone" (3 bks.) in which the motivation of the marriage has a different meaning. Rather than having Hades abducting or being given Persephone in an arranged marriage, Zeus granted her as a bride because Hades, frustrated by a wifeless and a childless state, threatened war on heaven.[80] Persephone's body was the means to prevent war between the upper and lower worlds; thus her virginity was sacrificed for Hades' use in return for harmony among the gods.[81] All the more meaningful is Zeus's admonition for her not to eat anything and, *au contraire*, she did eat and it was the pomegranate, a contraceptive. In Claudianus's poem, the peace would be threatened because Hades would be childless. How much of this was Claudianus's poetic license, and how much authenticity was there to the myth's early meaning?

The *Hymn to Demeter* can be seen as the taking of a story well-known by generations of Greek-speaking peoples and, before them, multilanguage peoples. Assuming the correctness of *Hymn,* composed around the seventh century BCE, the period reflects the formation of the city-state (*polis*) as tribal cohesion gave way to a larger political entity. The city-state society was concerned with marriage, child legitimacy, and a patriarchal culture that had emerged from one focused on a fertility goddess.[82] Zeus dominates over Demeter and Persephone. On the other hand, Claudianus lived during a period of the barbarian migrations and assaults on the Roman Empire, when rape was a reality. Likely, it seems to me, he picked up on an earlier theme in the Persephone story, where the issue was the eating of pomegranate in disobedience to the god (Zeus/Jupiter) in order to prevent conception and protect herself against giving birth to a child conceived through rape or forced marriage.

THESMOPHORIA: WOMEN'S SECRET INITIATION

A society known as the Thesmophoria was complete with secrets given to the initiates. Only Greek women could be initiated into its mysteries, as they would either camp out in tents some distance from town or, in some cases, on an island. At the Thesmophoria festival, the goddess honored was Demeter, but Persephone was also a part. Although few details are found in the ancient works (written by men, of course), various writers related incidental details. Aristophanes, the writer of comedy, wrote an entire play called *The Thesmophoriazuzae*, the name of a judicious council, held on the festival's third day. The plot of the play—to me Aristophanes funniest—was a trial of Euripides, the playwright, who was on trial *in absentia* for his misogynist portrayal of women; the indictment said Euripides described women as "double-dealers, false, faithless, tippling, mischief-making gossips, a rotten set, a misery to men."[83] So negative were women characterized that their husbands returned from the theater so sour that they peeped into closets to find lovers. As a male, Euripides could not defend himself so he pressed his father-in-law to dress as a clandestine woman to defend him. As the ploy unfolds, the man's presence was discovered with hilarious, bawdy results. In another source a man who complained that attendance at the Thesmophoria had a fee and he had to pay twice, once for his wife and again for his mistress![84] Despite the masculine humor about the activities of women on their female-only outings, the religious purpose was paramount for the attendees.

In the initiation ritual Persephone was first seen picking wild flowers. Likely there was a dramatization of the abduction, rape, eating of pomegranates seeds, and return to earth (even though there also was the tradition that she remained as the Queen of the Underworld, similar to Inanna's sister). Persephone left the earth as a girl and was raped most likely as a bride. Women who attended were usually married, although unmarried women as well could be initiated into the mysteries, called the Eleusian Mysteries because Persephone reportedly returned to the earth at Eleusis. (Another tradition said Sicily was the location where Persephone returned, and there also the Mysteries' secret ceremonies were performed.) The early Christian writer, Clement of Alexandria, who had no trepidations about exposing a pagan ritual, related some details among which was that women celebrants took care not to eat pomegranate seeds as they lay upon the ground.[85] Prytz Johansen studied the fragments of information about the Thesmophoria and noted the similarity with the Sacred Marriage motif that he regarded as the promotion of fertility.[86] Just as on the Uruk Vase, the women attending the Thesmophoria were there to celebrate both their fertility and, with the pomegranate, how to control it. Persephone moved from adolescent virgin to young woman and, perhaps, to motherhood. The girl in her was lost but she gained the knowledge of pomegranates that made Hell (Hades) more endurable for her stays there.

The agricultural fertility aspect was not lost. By reason of Persephone's disobedience (albeit in the *Hymn to Demeter* version because of trickery), from that time forth she had to return for some months each year to the Underworld and, when she did, the leaves turned, plants died, and, in Hesiod's description: "Avoid the month Lenaeon [late January, early February], wretched days, all of them fit to skin an ox, and the frosts which are cruel when Boreas blows over the earth."[87] That wintry months varied according to the number of seeds eaten from year to year appeared to be a reasonable deduction, but there is nothing explicit in the legend.

Ovid's *Fasti* links the three seeds to half the number of months that Persephone was allowed in the upper world.[88] In comparison to Persephone, Inanna secured her return to the Upper World by leaving behind her sister Ereshkigal. The god of wisdom joined Inanna, it will be remembered, in the Under World, paralleling the Adam-Eve story.

THE DEMETER VASE

The connection between Inanna and Persephone/Demeter is clearly seen on a vase, the Demeter Vase. A picture drawn on a Boeotian plate (see figure 2.1) shows either Demeter or Persephone (some say both in one) seated, enthroned and heavily draped. The figure's right hand holds a torch; her left holds two stalks of grain and two twigs with a pomegranate on each end. The torch in her right hand is, in Demeter's words to Persephone (placed there by Claudianus): "I had thought—as every mother does—of the wedding torch I'd hold to light your way from a happy nuptial feast to the bridal chamber and bed and the start of a decent life."[89] With her left hand, she holds the grain and pomegranates on an *omphalos* altar on which there is what appears to be a pomegranate.[90] The goddess holds the symbols for fertility and sterility. She controls each in her hand. Demeter's symbols are clear.[91] The juxtaposition of grain and pomegranates on the Demeter Vase is exactly the same as that on the Uruk Vase; the Demeter vase was almost 3,000 years after the Uruk Vase.

Figure 2.1 Boeotia plate (5th c. BCE) showing Demeter (possibly Persephone) seated holding torch in right hand and in left grain and pomegranates.

Source: Courtesy of National Archaeological Museum, Athens, Prot. No. 3579.

Graeco-Roman Medical Sources for Pomegranate as Contraceptive

Living in the time of the Emperors Trajan and Hadrian (98–138 CE), Soranus wrote in Greek a work, *Gynecology*, in which he listed some six contraceptive recipes for pomegranate:

> We shall make specific mention of some ["aid in preventing conception" 1:61]...
> Grind the inside of fresh pomegranate peel with water, and apply [as pessary].
> Or: Grind two parts of pomegranate peel and one part of oak galls, for small suppositories and insert after the cessation of menstruation.
> Or: Moist alum, the inside of pomegranate rind, mix with water, and apply with wool.
> Or: Of unripe oak galls, of the inside of pomegranate peel, of ginger, of each 2 drachms, mould it with wine to the size of a vetch pea and dry indoors and give before coitus, to be applied as a vaginal suppository...
> Or: Apply pomegranate peel with an equal amount of gum and an equal amount of oil of roses.
> Then one should always follow with a drink of honey water. But one should beware of things which are very pungent, because of ulcerations arising from them. And we use all these things after the end of menstruation.[92]

The remarkable observation is not that Soranus named pomegranates as a contraceptive but that there were writers who did not. Such medical authorities as Hippocrates (i.e., the various treatises known as the Hippocratic Corpus, *ca.* 440–330 BCE), Dioscorides (d. *ca.* 70 CE), and Galen (d. post 205 CE), gave detailed prescriptions about contraceptives but failed to mention pomegranates as having this power.[93] Similarly, neither was pomegranate recommended as a contraceptive in the medical works by Scribonius Largus, Oribasius, Marcellus Empiricus, Quintus Serenus, and Theodorus Priscianus, or the medical content in the encyclopedic work by Pliny the Elder.[94] Why should pomegranates have figured so prominently in the early history of West Asia and not be given for contraception in the Graeco-Roman period (Soranus excepted) nor in the Middle Ages (except for other medicinal uses)?[95] The reasons, I submit, are twofold: one, the pomegranate over the centuries, millennia really, had been horticulturally bred for tastes and the active estrogene compounds reduced. The wild pomegranate trees are found in the south Caspian area, southern Balkans, and northeastern Turkey. The pomegranate was bred by clonal propagation, not sexual reproduction.[96] Second, more effective contraceptive drugs were discovered, such as the famed silphium found only around Cyrene in North Africa, described by Juvenal as a "sure-fire contraceptive."[97] Cyrene's coins prominently displayed silphium, often with a woman touching the plant with one arm and her womb with the other. It resisted cultivation and became extinct by the fourth century CE—hunted out of existence by human exploitation.

Why not Egypt?

As I have mulled over the meaning of pomegranates for years now, I have been puzzled about the absence of pomegranates as a contraceptive in ancient Egypt.

There are two major sources for pharmaceutical data that are preserved on ancient Egyptian papyri: the Kahun Medical Gynecology (or Obstetrical) Papyrus (dating from around 1850 BCE) and the more famous and extensive Ebers Papyrus (dating around 1551–1500 BCE). Both are thought to be copies of earlier compilations with the Ebers said to date back to the Early Kingdom (as early as 3000 BCE). Major authorities agree that the pomegranate was not introduced into Egypt until the New Kingdom; the plant was brought to Egypt after the Syrian expeditions of Thothmes I (1525–1512 BCE).[98] The tomb of Hatshepsut (*ca.* 1470 BCE) held a large, dried pomegranate.[99] Some nineteen votive pomegranates were found in the tomb of Amenhotep II (r. 1450–1415 BCE), and Egypt became a major manufacturing center for pomegranate vases and jewelry.[100] Found in Tutankhamun's tomb was a rare silver vase shaped like a pomegranate, rare because silver is not often found in Egypt.[101] Once the pomegranate plant came to Egypt, its popularity soared.

Both the Kahun and Ebers papyri refer to contraceptives, mostly as pessaries; modern science and medicinal chemistry find virtually all effective to some degree.[102] (The only ingredient that was not is crocodile dung but, inasmuch as the crocodile represented the abortion god, perhaps the meaning was more religious than pharmaceutical.) Given the chronology, one would not expect to find pomegranates mentioned at all, but they appear in the Ebers Papyrus in treatments for dysentery, diarrhea, stomachache, and killing roundworms in the digestive tract.[103] There is a correlation that I have noticed between anthelmintic (worm killing) actions and contraceptive and/or early abortifacient drugs (to be seen as well in subsequent chapters). For example, Dioscorides prescribed pomegranate to kill intestinal worms but omitted its contraceptive quality.[104] (Moreover, in following his arrangement of drugs by affinities so that one could substitute for another, Dioscorides placed together pomegranates and date palms.)[105] Either the calendar date is incorrect for the introduction of the pomegranate in Egypt or the date for the papyrus copy of the Ebers Papyrus is misjudged.[106] In any case, however, the fact remains that there is no evidence that the ancient dynastic Egyptians employed pomegranates as contraceptives. Probably there were two factors at work: one was that the Egyptians already had what they judged to be effective contraceptives. The Hippocratic Aphorism Number One stated: "Experiment is dangerous."[107] Whatever the correct date for the pomegranate's introduction into Egypt and its recording in the Ebers Papyrus, the pomegranate would have been a recent plant in the ancient Egyptians' diet and pharmacy. According to my hypothesis, the pomegranate's active phytoestrogenic compounds were diminished by the time of the New Kingdom, or around 2,000 years after the story of Inanna and the pomegranate tree, the *huluppu*. By the time of the Egyptian New Kingdom, the pomegranate was unlikely to have been the contraceptive of choice.

Contraception and Contraceptives in Ancient Judaism

Among the Jews, just as in other cultures of the eastern Mediterranean, the guiding principle was to guard the male head of the family's right to have children. There was God's injunction: "increase and multiply; fill the earth" (Gen. 1:27–28). Some Rabbinic interpretations allowed its imperative as applying only to men.[108] The

Mishnah added a broader perspective: if a man had already sired a number of children (two males were mentioned as an example), he was excused from more propagation.[109] For example, a woman whose husband was killed or died could abort if it was her desire. A Talmudic statement gave circumstances in which a woman could employ contraception: a minor (where childbearing would be dangerous) and a woman whose pregnancy too closely followed a birth and jeopardized nursing of the infant. Rabbis differed as to whether these circumstances were limited to those outlined or, as examples, could be expanded to cover other circumstances.[110] Men who, it appears, knew little about the mechanisms of birth control wrote these sources. Specifically mentioned were a pessary (*mokh* in Hebrew), a "bitter medicine" (by implication orally taken), "cup of roots," and a "cup of barrenness."[111] Pomegranate could have been used, especially suggestive as a pessary, and for the bitter medicines, both willow bark and artemisia are likely candidates, both being effective and both well described as being bitter. The date palm was likely one of the birthcontrol measures. It is not possible, given the sources, to know precisely what medicines they took. What is certain is that Jewish women knew about contraception and abortion and, in some circumstances, were allowed by their husbands to use them. What is more likely is that some women used these measures without their husbands' knowledge.

What Was the Sin that Adam and Eve Made?

Today's woman takes estrogen to prevent pregnancy or, wishing the same and needing supervised medical intervention, she takes estrogen. Thereby, we should not call Inanna, Ištar, Demeter, and the other similar goddesses "fertility goddesses" but fertility-control goddesses. If my interpretation is reasonable, just like Inanna, she brought the control of fertility; so did Eve when she discovered the fruit of the tree of knowledge. Adam may have had no clue other than that he had disobeyed a godly order. This is, of course, a story, there being no "real" Adam—or Eve.

Genesis author "J" knew surely that pomegranates were contraceptives—the knowledge was pervasive in his culture. Without knowing that the fruit was the pomegranate, Merlin Stone, who studied early fertility goddesses, concluded about the sin: it was the "eating of the tree that gave her [Eve] the understanding of what 'only the gods knew'—the secret of sex—*how* to create life."[112] In telling the Garden of Eden story with the Fall of Adam and Eve, author "J" could not have had multigods interacting, because he believed in a single, all-powerful God. In developing his story, he would have used culturally derived stories. His theme is to account for humans' innocence in the Garden and subsequent disobedience and, with it, how immortality was no longer possible, shame held in nakedness, and sexuality leading to the destiny of women to endure pain and danger in childbirth. There is nothing in his telling of the story that indicates that God's displeasure over the eating of the pomegranate was contraception per se. After the Fall, humans were responsible beings and had to account to God for their actions. Responsibility came with the roles that sexuality assumed in human society and, by deciding when conception was to occur; humans' relationship to God was altered just as Inanna's fulfillment and power came from her use of the pomegranate. Inanna was reborn, at a price (leaving her sister in the Underworld); Eve and Adam paid a price for disobedience: the price was responsibility. There is no hint in any of the ancient stories, including the one in Genesis, that contraception was wrong, but it represented empowerment through knowledge. Responsibility is imposed on humans because they disobeyed God just as Inanna disobeyed her divine family's wishes that she not go to the Underworld to receive the pomegranate, and Persephone disobeyed by partaking of a contraceptive to thwart godly injunctions. The ancient Israelites did not select the pomegranate as their symbol because of its taste; they chose it because the pomegranate changed the relationship between their God and humans, and, by inference, in a positive way. The pomegranate changed the way humans thought and acted about sex and empowered them with knowledge to control their destinies. The "fall" was not so much of a fall as it was, in the words of John Evans, the biblical scholar: "an over-abrupt rise, a momentary transition from a state of naïve innocence to one of precocious maturity."[113] Adam became a man and Eve a woman. The serpent was necessary because the "J" author knew that people knew the Gilgamesh story about the serpent stealing the plant of life and probably the serpent in the pomegranate roots from the Inanna story. The transition to maturity and responsibility carried with it the propensity to sin. The pomegranate did more than symbolize the transition; it was the message for responsible reproduction and human conduct.

An Epilogue on Pomegranates

A male person walking beneath the pomegranates on top of the two columns before the temple door in Jerusalem about the time of Augustus Caesar probably would not have been aware that pomegranates were contraceptives, since, by that time, their use had been reduced to quaint and largely forgotten lore. The pomegranate was prominently displayed in Greek, Etruscan, and Roman mythology and even appears in early Christian iconography. A vase from Campania has a love scene with a man and woman on a couch while, within arm's reach, is a pomegranate on a table (see figure 2.2).[114] Among the many representations in frescoes are the Aphrodite and Adonis painting from the Casa di Adone in Pompeii (in 79 CE) that has three pomegranates on an altar.[115] Finally, a floor mosaic from the Villa by Hinton St. Mary, Dorset (4th c.), has Christ with pomegranates on each side.[116] We know the mosaic is of Christ, because on his left and right appear, respectively, the Greek letters "Chi" and "Rho," the first two Greek letters for "Christos/Χριστός." In early Christianity, the pomegranate became a symbol for resurrection. Christ associated with an antifertility drug is startling at first, but Christians supplied revised meaning: the pomegranate assimilates Christ's descent and resurrection and connects with Persephone's rebirth on Earth and a return to things green and fertile. The Romans of the Empire continued to use contraception as well as early-term abortifacients, but no longer were the agents pomegranates.

The pomegranate's tough shell surrounding many seeds allegorically symbolized many Christians under one authority, the Church. Finally the Christians extended the pomegranate to symbolize the Virgin's chastity.[117] Sculptor Jacopo Della Quercia (1374–1438) executed his famous Madonna of the Pomegranate.[118] Holy Roman Emperor Maximilian I (1459–1519) and Isabella of Portugal, wife of his grandson,

Figure 2.2 Roman love scene with pomegranates on stand within easy reach.
Source: Courtesy of Martin von Wagner Museum, Antikenabteilung, Universität, Würzburg, Germany.

employed the pomegranate as their *impresa*.[119] What people in the early modern period regarded as the meaning of a symbol may not have been the same meaning it had in the ancient world. Still, the ancient meaning of the pomegranate as life-and-death, fertility and infertility, is linked to the medieval and early-modern meanings.[120]

An Epilogue on Eve

The story of Adam and Eve, Eve in particular, did not play a large role in early Jewish thought and letters, it appears. In early Judaism, Eve was "more humorous or mischievous than vicious," or so said John Philips,[121] but to modern conventions, she was weak and twice-blamed: first, for listening to the serpent in disobedience and then for persuading Adam to disobey by eating the forbidden fruit. What the Talmudic Rabbis took from the message was God's injunction to be fruitful and multiply, but the Mishnah portrays a woman as a "sexual volcano always on the brink of eruption," in the words of Pamela Norris, who wrote a "biography" of Eve.[122] St. Paul made Eve more of a central figure in Christianity. First (2 Cor. 11:3), he wrote: "But I am afraid that as the serpent deceived Eve by his cunning, your thought will be led astray"; second (1 Tim 2:13–14), Paul betrays his misogyny (*viz.* 1 Cor 6:15–20; 11:3–16) by outlining why a woman should be silent and submissive: "For Adam was formed first, then Eve; and Adam was not deceived, but the woman was deceived and became a transgressor."

The Christian Church Fathers used the "J" section of Genesis to account for original sin,[123] a doctrine, which in Judaism, if present, was, at least, submerged. Primarily they interpreted the sin was sexual misconduct, with the disobedience in eating the forbidden fruit. Poor Eve, the "mother of all living," was once a supreme goddess, now relegated to mere humanity, taken from a man's rib in a true patriarchal society, made a seducer, and, with Christianity, designated the first sinner and responsible for humanity's downfall. By the sixth century, Mary, the silent, suffering virgin, was a counter to negative female imagery, during the early medieval period, to treacherous Eve. But, after the Romance poets developed the notion of chivalry, the scholastics in the medieval university reinterpreted Genesis' Garden of Eden story to make the point that Eve, being the last to be created, was the superior creature. The world has turned many times since the thirteenth century, when scholastics tried to rescue Eve and place her on a pedestal, not on a stool in the corner. Eve has come a long way since she was named Inanna, who had discovered the qualities of the pomegranate.

Chapter 3

Mandrake, the Love Apple, and the World's Religions

> Genesis 30: 14–17: In the days of wheat harvest Reuben went and found mandrakes in the field, and brought them to his mother Leah. Then Rachel [Leah's younger sister] said to Leah, "Give me, I pray, some of your son's mandrakes." But she said to her, "Is it a small matter that you have taken away my husband [Jacob]? Would you take away my son's mandrakes also?" Rachel said, "Then he may lie with you tonight for your son's mandrakes." When Jacob came from the field in the evening, Leah went out to meet him, and said, "You must come in to me; for I have hired you with my son's mandrakes." So he lay with her that night. And God hearkened to Leah, and she conceived and bore Jacob a sixth son. [Rev. Standard trans.]

The mandrake enabled Leah to conceive two more sons, Issachar and Zebulon, and a daughter, Dinah. There is more to the story. Leah felt displaced because Jacob loved more Rachel, Leah's younger sister, and Leah had become infertile. Before Reuben's mandrake, Rachel, who had always been infertile, allowed her handmaiden to lie with Jacob and to produce two sons by Rachel's proxy. The mandrake altered the situation in a dramatic way. Rachel permitted Leah to sleep with Jacob provided she also received the mandrake as a result of which Rachel had two sons, Joseph and Benjamin. Leah wanted the plant so that she could resume fertility.

The fertility plant enabled barren Leah to add to the previous sons: Reuben, Simeon, Levi, and Judah (Gen. 29:31–35–30:1–25). Rabbi Levi in the *Midrash Rabbah* exclaimed: "Come and see how acceptable was the mediation of the mandrakes, for through these mandrakes there arose two great tribes in Israel, Issachar and Zebulon"[1]—pointing only to Leah's sons. Although Leah was the eldest and Jacob's first wife, he had also married her younger sister, Rachel. According to a Jewish legend, Leah said: "God promised Jacob twelve sons. I bore him six, and each of two handmaids has borne him two. If, now I were to bring forth another son, my sister Rachel would not be equal even unto the handmaids."[2] Each of Leah's six sons was a primogenitor for the twelve tribes of Israel—the others coming from

Rachel and the two handmaidens. Judah was the most prominent because from him came the kings of Israel, including David, and from the house of David came Jesus. In time, Judah's kingdom was called Judea and its people Jews by encompassing the other eleven tribes. Leah would not have had two of her six sons and Rachel her two sons, Joseph and Benjamin, had not the mandrake "secured" four sons.[3] Had Leah not resumed her fertility, she would have lost status and, with it, her sons might have lost favor. The mandrake plant was responsible for four of Israel's tribes.[4] Later Jewish and Christian theologians recognize the importance of the mandrake, but only in an incidental way. The mandrake's role was not incidental but physiologically central to how the ancient tribes of Israel came into existence and how the Christians trace their Messiah's lineage. Certainly Reuben's finding of the mandrakes was of great importance to Leah and Rachel. John Skinner, a biblical scholar, believes that the "incident of the love-apples [mandrakes] is a piece of folklore... [that] the writer (J) [of the Genesis section] curtailed." Skinner adds, "The story must have gone on to tell how Rachel partook of the fruit and in consequence became pregnant, while Leah also conceived."[5] Why would the author include the mandrakes? They were essential to the story.

Harry Potter

J. K. Rowling is another authority on the mandrake—albeit, less so than the Bible; she is the author of *Harry Potter and the Chamber of Secrets*. Harry and his classmates are ushered into Greenhouse #3 where an earmuff was placed before each person. Professor Sprout asks who can "tell the properties of the Mandrake?" Harry's friend Hermione answers: "Mandrake, or Mandragora, is a powerful restorative.... It is used to return people who have been transfigured or cursed to their original state." But it is dangerous, Sprout adds and he asks why. Again, Hermione: "The cry of the Mandrake is fatal to anyone who hears it." Correct, again, replies the professor and she adds that their greenhouse mandrakes were "only seedlings," whose cry was too weak to be fatal but they "*will* knock you out for several hours."[6] The earmuffs prevented the Hogwarts students from passing out as the baby roots screamed.

Between Genesis, Shakespeare, and Harry Potter

A mandrake's cry, potentially fatal, as it is pulled from the ground was not imagination in Rowling's Harry Potter stories; the crying of the mandrake root is older than Harry. William Shakespeare's *Romeo and Juliet* (Act 4.Scene 3) reads: "And shrieks like mandrakes' torn out of the earth." Shakespeare also has the Duke of Suffolk say: "Would curses kill, as doth the mandrake's groan" (Henry VI, Pt. 2. Act 3. Sc. 2). Many are the legends surrounding the mandrake, some of which we shall visit, and there is also a large amount of medical lore about one of history's most famous medical plants. The questions are as follows: (1) Does the mandrake have the qualities that history has attributed to the herb; (2) How has this plant worked in history to change human society, both in legend and in pharmaceutical action?

THE BIBLE

When I took a course in the Bible in a public school, we had to memorize all those "begets," who married whom and who were their children. At the time my thoughts were "what a waste of time—just like algebra—and I'll never need to know this stuff." In Genesis (13: 15–16), God promised Abraham, the patriarch, a blessing involving land and his progeny, but he had to leave his land for an unknown land and for an undisclosed length of time. In the wilderness, he was unable to bear a child through Sarah, his wife, who, in accordance with custom, arranged a concubine in Hagar, an Egyptian slave. An angel told Hagar that she would have Abraham's son and call him Ishmael. Sarah's duty to ensure her husband children soured her relationship with Hagar who was banished to the desert wilderness. God told Sarah she too would conceive, but, knowing her old age, she laughed "to herself" (Gen. 18:12). God told Abraham that Sarah would bear his future child and "she will be a mother of nations; kings of peoples shall come from her" (Gen. 17:16). Sarah's son was Isaac, whose son was Jacob. Thus, the Jewish people are regarded as the descendants of Jacob. Ishmael, on the other hand, was the primogenitor of the Islamic peoples, and about whom much more is told in the Qur'an.[7] Three of the world's major religions hinged on female fertility; all trace their ancestry back to Hagar and to Sarah, Rebekah, Rachel, and Leah. Sarah's son was conceived through divine intervention but, in Leah's case, it was through the power of the mandrake plant, with the implication of God's approval. Although not so stated, the miracle of Sarah's fertility may have been also the mandrake, should God be perceived as working through natural processes as he did with Jacob's wives.

FERTILITY AND DIVINE FAVOR

Fertility was very important to some societies in the ancient and modern West Asian region during periods of time. In his despair, Abraham cried: "O Lord God, what wilt Thou give me, for I continue childless" (Gen. 15:2). Second only to death, a major divine punishment was the curse of sterility (Lev. 20: 20–21). The *Genesis Rabbah* (a commentary on Genesis, probably written slightly later than the two Talmudic Rabbi accounts) accepts that conception was a sign of divine favor (although some opinions differed on Hagar).[8] Hagar was a household slave until she became pregnant; then she received direct divine communication from God through an angel. Sarah's conception in her old age was an indication of God's direct intervention to fulfill his promise to Abraham.[9] Amazingly the four Jewish matriarchs all had fertility problems: Sarah, Rebekah (wife of Isaac and mother of Jacob, conceived after twenty years of marriage), Leah (Jacob's first wife, who became barren after delivering four sons), and Rachel (second wife of Jacob and mother of Joseph and Benjamin). A married but barren woman was not only regarded as being undesirable but her condition rose to the level of legitimate divorce.[10]

Relative to Rachel's fertility after eating the mandrake, the Midrash Rabbah asserted: "Were she not righteous, would she have borne children?"[11] Twentieth-century Muslim women of Jordan either tied a mandrake around their necks and/or ate mandrake when they wanted to become pregnant. Raphael Patai, who studied mid-twentieth century Muslim society in modern Israel and Jordan, said: "Barrenness is still considered the greatest misfortune and disgrace that can befall a woman."[12] It appears incongruous that three generations of women involving four women would be afflicted with infertility. Sarah was too old for childbearing, and she was anxious about being replaced by a new wife for Abraham.[13] Certainly all these matriarchs would have had one condition in common: stress. This is where the mandrake reenters the discussion.

The Mandrake Plant

Botanically the mandrake is a member of the Solanaceae family, one famous, even notorious, in herbal lore because of its heavy constituent tropane alkaloids, which include atropine, hyoscymine, and scopolamine. Plants in the same family include the potato, tobacco, and deadly nightshade (also called belladonna). Botanical references occur under several binominals: *Mandragora officinarum* L., which comprises two distinct species: *Mandragora autumnalis* Bertol. and *Mandragora vernalis* Bertol. or, alternatively *Atropa mandragora* L. I shall use the preferred *Atropa* name for the genus. As we shall see below, many ancient authors say that there are two "kinds," male and female. Whereas we might speculate about whether the gender differentiations are botanically based or appearance-based on the root shape, the assumption that I shall make in this study is there is no practical difference from the standpoint of the physiological effects of either kind. This assumption is based on a chemical study that shows that the variations between the two species "were no more than would be expected as a result of normal biological variation."[14]

The American mandrake, also called mayapple, is a different plant: *Podophyllum pellatum* L., of the Berberidaceae family, which also has strong alkaloid compounds similar but not the same in their physiological effects. Confusion may arise from the English common name, devil's apple, as applied to both the European and American mandrakes. The mayapple is a common garden ornamental; no one should ingest any part of the plant because, although used medicinally, it is highly poisonous—and so is the European mandrake. The European mandrake's habitat is the Mediterranean region. Late medieval efforts to grow it in greenhouse conditions as far north as Britain were successful. Grieve's *Herbal* has this description of it:

> It has a large, brown root, somewhat like a parsnip, running 3 or 4 feet deep into the ground, sometimes single and sometimes divided into two or three branches. Immediately from the crown of the root arise several large, dark-green leaves, which at first stand erect, but when grown to full size (a foot or more in length and 4 or 5 inches in width) spread open and lie upon the ground. They are sharp-pointed at the apex and of a foetid odour. From among these leaves spring the flowers, each on a separate foot-stalk, 3 or 4 inches high. They are somewhat of the shape and size of a primrose, the corolla bell-shaped, cut into five spreading segments, of a whitish colour, somewhat tinged with purple. They are succeeded by a smooth, round fruit, about as large as a small apple, of a deep yellow colour when ripe, full of pulp and with a strong, apple-like scent.[15]

The plant was found mostly in shaded woodland areas with plenty of moisture and good drainage and loose sandy or loamy soil. In antiquity, the plant resisted cultivation, which, although tricky, could be done from seed. The greenhouse scene in Harry Potter showed the plants with roots, looking like babies, as they were being noisily transplanted. The roots are similar to a carrot (Grieves said parsnip) but different by virtue of two major roots coming down from the top and two from the bottom. In this way the root systems resemble a human form.

EGYPTIAN LORE

Elephantine was an island in the Nile at the First Cataract (present-day Aswan) and known as the place where mandrake beer was manufactured. A nineteenth-century Egyptologist, Gaston Maspero, said that the mandrake drink was "employed either in medicine or in magic."[16] This mandrake drink figured prominently in Egyptian religion. Aspects of the myth are familiar because of similarities with biblical and Mesopotamian stories regarding the near-destruction of humans who had become estranged and unmindful of God (or the gods). Egypt's chief deity, Ra, sent Hathor (said in some legends to be his wife and mother of Horus) to slay all people, a task which she readily performed. Gods cannot err or, at least, readily admit to it. He regretted his order and asked Hathor if her enthusiasm might be excessive. Hathor replied: "By thy life when I slaughter men then is my heart right joyful!" Ra had a miracle to save humans: the mandrake! Sending to Elephantine for a large supply, he called upon a miller to crush the grain, brew beer, and mingle it with human blood from those slain. While Hathor slept, Ra ordered a courier to carry 7,000 jars with the specially manufactured mandrake beer and, by emptying the jars, to cover the ground four-hands deep. Awakening to resume her carnage, she drank the beer tasting of human blood and mellowed. Hathor fell into a drunken stupor. Upon awakening sober, she had a relaxed feeling and, with it, a changed heart; she ceased killing humans. Thus was humankind saved. Ra ordered that henceforth on Hathor's feast day, all humans should partake of mandrake beer and sleep thereafter.[17]

Although represented many ways, Hathor was seen as a pregnant hippopotamus with a human or hippo head and a charmingly swollen womb from pregnancy.[18] A picture from the Book of Dead (19th Dynasty) shows the hippopotamus deity, Taweret, standing before Hathor with a hippopotamus head.[19] Among women, Hathor was probably the most beloved of the gods and, so it appears, Hathor and Taweret became one in Egyptian women's perception. One of Hathor's names was "mistress of the vagina": she was associated with conception, pregnancy, birth, motherhood, and all aspects of loving child rearing.[20]

MARRYING HATHOR TO BIBLICAL MANDRAKE

Finding the mandrake as a fertility agent in Egyptian medicine is expected but, alas, it is not to be found, despite some erroneous efforts to say otherwise.[21] According to an authority on Egyptian herbs, Lise Manniche, the mandrake is found in no pharaonic texts, including the famous Ebers prescription papyrus, despite the fact that this compendium has approximately 1,000 recipes. As noted in chapter two of this work, however, many, some say all, recipes date back to the Old Kingdom. *Rmmt* is the Egyptian word for mandrake; pictures of the plant are frequently found on tombs, monuments, and artifacts from the New Kingdom, the period when Manniche says it was introduced into Egypt. Its New Kingdom representations gave it a "symbolic erotic significance in pharaonic times," says Manniche.[22] The mandrake (*rmmt*) is often mentioned in Egyptian love poetry.[23]

My study of ancient Egyptian papyri (which included the Kahnun Obstetrical or Gynecological Papyrus, *ca.* 1850 BCE) found no mandrakes.[24] Yet, the Egyptians knew the mandrake from at least the New Kingdom, similar in that respect to the pomegranate. Mandrake representations appear in tomb paintings and depict the plant with its apple-like fruit.[25] A carved ivory box-lid shows the pharaoh's daughter picking mandrake apple in Tutankhamun and Queen Ankhesenamun's garden.[26] Curiously in another scene the Queen holds two mandrake fruits near the nose of the Pharaoh, causing one Egyptologist to exclaim: "Does it suggest that she intended to give potency to the rather limpid boy with a belly in the Akhenaten-tradition?"[27] No depiction, however, that I have seen shows its roots from whence comes its strongest medicinal compounds, as Harry Potter's friend knew.

The Love Apple as the Mandrake

For once among scholars, there is virtually no controversy about what was the "love-apple" in Genesis 30:14, which Reuben found in the fields; Genesis' love apple is Harry Potter's and our mandrake. The transliterated Hebrew word in Genesis is *duda'im,* meaning "love apple" from root *dwd* ("to love" or "to fondle"). The Arabic word for mandrake, *yabruh,* also means "love apple."[28] The seventy or so Jewish scholars who translated Genesis into Greek in the Septuagint version in approximately the third century BCE, rather than translate the word to "love," employed the Greek name for the plant, *mēlon mandragora,* whose identification is certain. Moreover the Vulgate, Aramaic, Syriac, and Arabic versions all agree on the mandrake, whereas the rabbinical tracts come about as close to agreement as the Rabbis did on anything.[29] The Hebrew *Song of Songs* (or, *of Solomon* 7:13–14) in the context of love-poetry: "The mandrakes *(duda'im)* give forth fragrance, and over our doors are all choice fruits, new as well as old, which I have laid up for you, O my beloved." The authorities on biblical botany are in agreement about the "love-apple."[30] Moreover, the connection between the plant and legends about it as seen in discussions in the Talmud and Josephus, the Jewish historian (as we shall see below), make even more certain its identification because the legends have persisted for millennia. For example, an old German custom was to place mandrakes under the nuptial bed.[31]

Ugaritic texts containing one of their myths, *Ba'al et 'Anat,* has *ddym* being placed in the ground that some scholars connect with the Hebrew *duda'im* and translate a mandrake, but the context makes the citation uncertain.[32] Even more intriguing is Hesychius, a Byzantine lexicographer of fifth century CE, of Alexandria who gave the synonym of Aphrodite as being Mandragoritis, "Lady of the Mandrake."[33] Aphrodite, as we saw in chapter one, was the classical Greek goddess of sexuality and love.

MANDRAKES IN HOMER?

Frustration to an historian is not being able to put a date on something, especially something intriguing and important. From very early times—but how early?—the mandrake root was associated with a human form. An etymology for *mandragora* or mandrake is related to the Indo-European and Persian for *man giyā* meaning "Man Plant."[34] Columella, the Roman agriculture writer, called it "quasi-human/ *semihominis*,"[35] but its looks only contributed to the stories about it. What made mandrake legendary were its physiological effects on humans. Homer described the goddess Circe's use of a *pharmakon,* to cast spells over wayfarers on an island where she lived after poisoning her husband. Some of Odysseus's companions she turned into pigs. Odysseus escaped by using a plant called "*moly*"/ *mōlos* to protect himself and eventually was able to persuade her to restore his friends.[36] Columella and Pliny[37] say that the mandrake is also called "circaeon," doubtlessly after Circe who was a goddess of love (or degrading love) and associated with Isthar, the Assyrian Inanna counterpart. So, what plant was *moly*?

Jerry Stannard's careful study indicates a confusion of plants, even though traditionally *moly* is identified as the mandrake, partly on the basis of Dioscorides giving mandrake's synonym as *Circaia,* "the Circe plant."[38] Stannard believed that some of the classical citations, including Homer, communicated a description of *Withania somnifera* Dun., a member of the Solanaceae family.[39] Much of Stannard's evidence comes from the pharmacological actions it was alleged to have had. I suggest that *moly* was a species of mandrake, on the basis of Homer's description, albeit brief: "So saying, Argeïphontes gave me [Odysseus] the herb [*pharmakon*], drawing it from the ground, and showed me its nature. At the root it was black, but its flower was like milk. *Moly* the gods call it, and it is hard for mortal men to dig; with the gods all things are possible."[40] Whereas mandrake's root is usually described as being brown, it is a dark brown; its flower's color varies from whitish to light violet and is aptly called milk-colored. Most of all, only mandrake has the legend surrounding the danger of digging it up. Finally, there is the synonym Dioscorides gave: "Circe's herb." Later the name shifted to other plants, just as Stannard documented. Dioscorides's *moly* was probably a member of the Allium (onion). Much to the frustration of botany historians, all too often plant names shift among species, families, and genera. Homer's *moly* was probably the mandrake, and, if not, it surely was a related species in the Solanaceae family as Stannard proposed.[41] Likely Homer's *moly* has the same physiological effect as the mandrake or one of its cousins. Inexplicably, however, Homer presented *moly* as an antidote to Circe's poisons. Again, religious stories make sense often in a murky way. Homer may have begun the Greek stories about the mandrake (or its related species), but the ancient Mesopotamian and Egyptian stories are clearer, as we shall see in the next section.

CLASSICAL AND MEDIEVAL MANDRAKE LEGENDS

There were many legends that ancients told about the mandrake, and there is no confusion about the plant's identification, inasmuch as *mandragora* is almost always the name given. Anne Van Arsdall succinctly summarized the combined ancient legends, as follows:

> The mandrake is a plant with a large root shaped like a human being, and it screams when pulled out of the ground. The scream kills anyone who hears it. So to safely pull up the mandrake root, dig down to expose the root using a tool of ivory (some say iron), making three circles about the root and chanting while digging. Then fasten a long rope around the exposed root. Tie a very hungry dog to the other end of the rope. You must be careful to cover your ears at this point. Take some meat and then throw it just out of the dog's reach, and he will lunge to get the meat, pulling the mandrake out of the ground and causing it to shriek. The dog will die because he hears the scream. You can then safely capture and use the mandrake root.[42]

Theophrastus related one of the earliest accounts by describing drug dealers and root gatherers (*pharmakopōlai* and *rhizotomoi*) who with a sword draw three circles around the mandrake, and cut it when they face westward as they dance around the plant chanting as "many things as possible about the mysteries of love."[43] The root gatherers (*rhizotomoi*) had a reputation for magic and use of mysticism and, given the comparative rarity of mandrake in the wild and its strong effects, it must have garnered a premium price. Whatever the origins of the mysteries regarding the mandrake, including death for man or dog who pulls it up, the story served well those who gathered it. Ordinary folk must have been frightened away. Anne Van Arsdall has added an intriguing hypothesis about the use of dogs in its harvesting. Many accounts speak of its smell. Perhaps dogs were trained to find it in the woods and wastelands.[44] Whereas the mandrake had precious little attention, given its pronounced qualities, in the ancient medical literature, its importance and association as a primary herb is seen in one of two frontispieces in the famous Dioscorides's manuscript known as the Juliana Anicia. These images feature the mandrake and signify the mandrakes' importance in herbal lore. Dioscorides, seated, points to the mandrake held by Epinoia, the personification of Thinking, while in the foreground is a dog. A similar image in an Arabic translation of the same text shows Dioscorides, seated, with a student while demonstrating with a mandrake. A medieval manuscript of Pseudo-Apuleius' Herbal shows a dog pulling up the mandrake (see figure 3.1). The medieval period intensified and supplemented the mandrake legends, as we shall see in the last chapter. Also, the Middle Ages transformed the plant into more of the devil's tool.

Sources quoted earlier, from Homer and Theophrastus to the Middle Ages, do not inform us about the details of the lost legends from ancient Mesopotamia, Syria, and Egypt (the Hathor legend excepted). A Jewish legend supplements Genesis on the mandrake. Reuben found the mandrake in a way that mirrors the mandrake legends. While he tended the fields, he tied his ass to a mandrake. The ass pulled up the mandrake by the roots and then died. Reuben found his ass dead and the mandrake, which he safely took to his mother.[45] Leah, it was said, acted on "pure, disinterested motives" to resume childbearing but it is difficult to reconcile the view with a statement she was alleged to have made. "If, now," Leah said, "I were to bring forth another son, my sister Rachel would not be equal even unto the handmaids."[46] Rachel was "not a pious woman" but "by bearing a son, she had escaped another

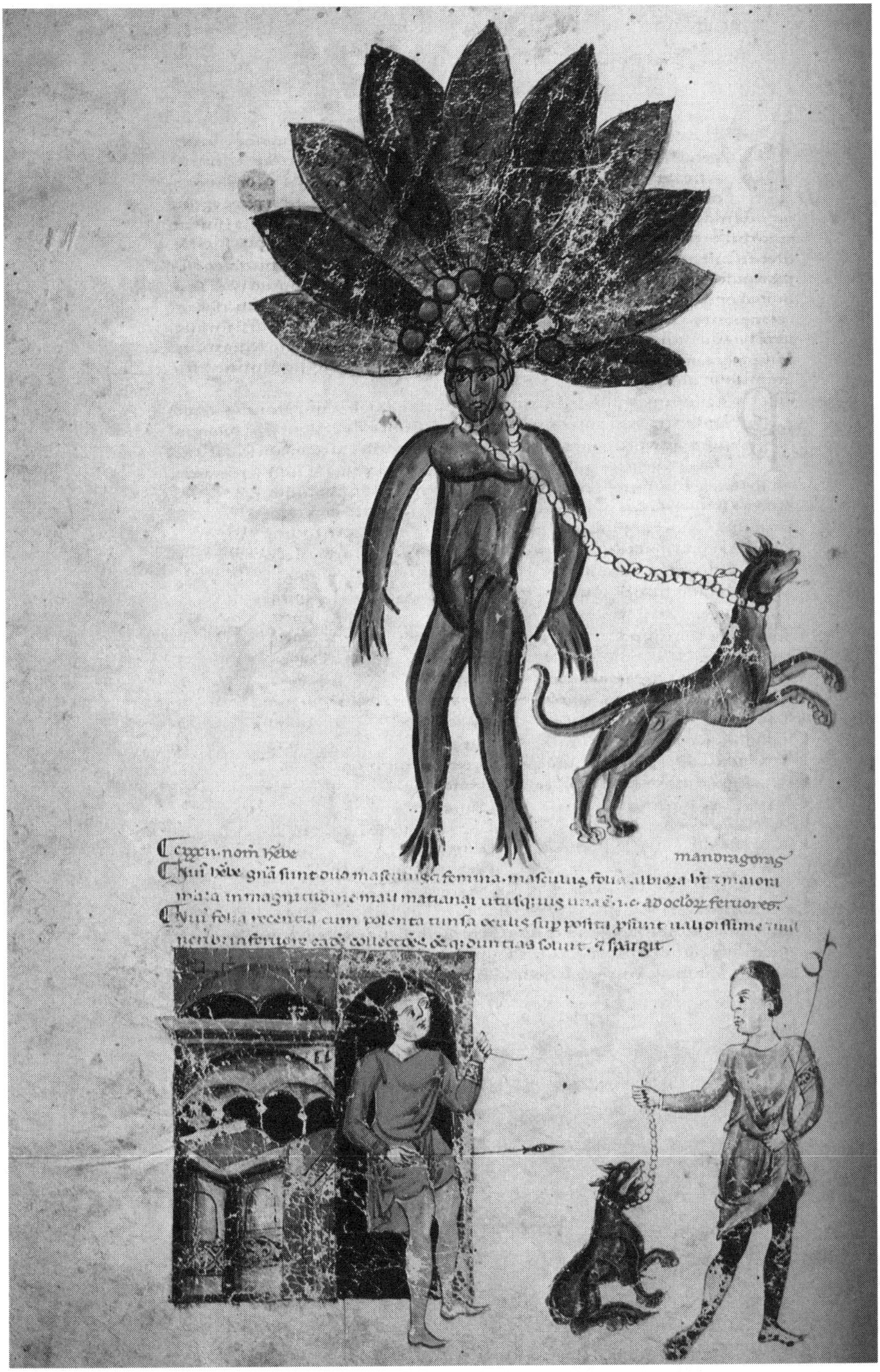

Figure 3.1 Mandrake uprooted by dog on chain.

Source: From Herbal of Pseudo-Apuleius, Vienna, Ms lat. 93, 13th c., fol. 117v. Courtesy of the Austrian Nationalbibliotek.

disgrace." Her second son, Benjamin, had ten sons, Joseph, the eldest, two sons, causing Rachel to proclaim that she had provided Israel's twelve tribes. Had she not been guilty of this prideful statement, she "herself would have begotten twelve tribes with Jacob," or so said the legend.

Josephus (37–post-100 CE), a Jewish historian, seems to be describing the mandrake as determined by context and so assumed by many scholars through the ages. He called it Baaras, a name taken from the town near the Dead Sea:

> In the ravine, which encloses the town on the north, there is a place called Baaras, which produces a root bearing the same name. Flame-coloured and towards evening emitting a brilliant light, it eludes the grasp of persons who approach with the intention of plucking it, as it shrinks up and can only be made to stand still by pouring upon it certain secretions of the human body.[47] Yet even then to touch it is fatal, unless one succeeds in carrying off the root itself, suspended from the hand. Another innocuous mode of capturing it is as follows: They dig all round it, leaving but a minute portion of the root covered; they then tie a dog to it, and the animal rushing to follow the person who tied him easily pulls it up, but instantly dies—a vicarious victim, as it were for him who intended to remove the plant, since after this none need fear to handle it. With all these attendant risks, it possesses one virtue for which it is prized; for the so-called demons—in other words, the spirits of wicked men which enter the living and kill them unless aid is forthcoming—are promptly expelled by this root, if merely applied to the patients.[48]

Josephus related a second story that says that he who pulls up the mandrake by the root promptly dies. To prevent a human death, a dog is tied to the roots, pulls up the plant, and dies.[49] These legends (ritual harvesting and death of dog) point to Baaras's identification as the mandrake, which many scholars believe.[50]

In Josephus's discussion of the fertility actions of Leah and Rachel, Josephus uses the Greek term *mandragorou mēlon* ("mandrake apples") to identify the plant.[51] The simplest explanation for the different terms is that, when Josephus was reporting on the plant near Baaras, he gave its local name, but, when he was identifying the plant in Genesis 30:14, he knew it to be the mandrake.[52] Josephus's knowledge about the plant would come from his Jewish culture.

The Meanings of Mandrakes in Mesopotamian Languages

The mandrake legends in the languages of ancient West Asia and Egypt are more difficult to ascertain. In Mesopotamia and in Semitic-speaking lands there is less certainty, although scholars are fairly definitive that mandrakes grew there—even in modern times—but exactly what were the words? R. Campbell Thompson identified mandrake as the meaning of the Assyrian word, *Nam.tar (ira)*, literally "male drug of Namtar," the devil god.[53] An Arabic word for mandrake is "devil's testicles," and, as already given, the common English name is "devil's apple." Similarly a German word for mandrake is *Zauberwurzel* ("sorcerers' root"). In the eleventh century, Hildegard of Bingen understood: "In *mandragora* [mandrake] the influence of the devil is more present than in other herbs."[54] In Germany, where Hildegard lived, a mandrake was placed under a bridal bed.[55]

Sumerian Love Apples

In the first chapter we saw the incantation to Inanna with these lines:

"Inanna, who loves apples (*giš.hašhur*) and pomegranates (*giš. nu.u'r.ma*), has brought forth potency."[56] Probably the word *giš.hašhur* is mandrake rather than our domestic apple as traditionally translated. *Giš* is the predeterminate meaning, what follows is a small tree. To repeat what was said in chapter one, our domestic apple trees do not grow in a hot climate, such as Mesopotamia, and there are no depictions of the apple in the art and artifacts of the land between the rivers. In the Inanna incantation, we have the juxtaposition, again, of a drug that stimulates love (i.e., an aphrodisiac) and fertility against one that prevents conception. Although Thompson and the Sumerian word-list identify *giš.hašhur* as the cypress or Persian oak tree or its resin ("abundance" + "to pour" + "tree trunk"),[57] I tentatively propose that the Sumerian love-apple plant is the mandrake.

Two Opposite Traditions

How do we reconcile the two traditions? On one hand, the mandrake is an invaluable aid in reversing infertility. In contrast, mandrake is associated with the Devil or Evil, going back to the male god, Namtar, the devil god. Mandrake is a good, benign, and blessed herb, according to one tradition and, according to another, a plant that works evil. In 1630, three women in Hamburg were sentenced to death for merely having mandrakes in their homes.[58] An answer to the dichotomy lies, I believe, in what the mandrake did as a drug.

Ancient Medicinal Actions

Whereas Egyptian medical sources, most of which date before the mandrake introduction into Egypt, are silent, the Assyrian texts are filled with mandrake in prescriptions, mostly as a narcotic and sedative. One recipe for headache, constipation, muscular inflammations, to name some symptoms, contains both mandrake and opium.[59] Recipes containing mandrake (usually specified as male mandrake) were given for difficulty in urination, male penis "dribbling," hemorrhoids (suppository), constipation (suppository), rectal disease, and general rectal and urinary problems.[60] Another recipe places the male mandrake (juice?) alone on a tooth that aches.[61] Two more recipes contain male mandrake for chapped lips, another recipe includes mandrake that is used for a variety of oral problems (some of which are lost in the text),[62] and there is another for excessive salivation.[63] Other recipes include the mandrake for liver complaints, difficulty in breathing, jaundice, and difficulty with child delivery.[64] The latter may appear as magic, because the text says to rub on the womb seven times the "root of Namtar-ira [*ira*=male] from the north."[65] The action is probably not magic but science: the medicinal action of a sedative-narcotic can be absorbed through the mucosa.

Mandrake's Medicinal Actions

Greek mythology and modern science intersect with the preferred Linnaean binominal name for mandrake: *Atropa mandragora*. Early in the cosmos' creation, the gods produced the three fates, to which all are bound: Clotho (a spinner who weaves the thread of life), Lachesis (chance), and Atropa (inescapable fate from which there is no appeal).[66] An example of Atropa's action in our time is genetics and in Hellenic times deadly poisons. One of the meanings for the Greek term *pharmakon* was poison, but the Greeks knew that some poisons taken correctly for the right purposes could act beneficially. Mandrake is an anticholinergic drug whose action is caused by occupying the muscarinic acetylcholine receptor. Today, almost anyone older than forty years had atropine—the chemical name coming from the fate Atropa—administered for ophthalmic use as a drop to dilate the eye. In present practices, it is less frequently used in ophthalmology. Among present uses are as a resuscitation for bradycardia (slow heart function often associated, in layman's terms, with a heart attack) because its action on the vagus nerve causes heart rate to increase.[67] Other plants containing atropine are belladonna (*Atropa belladonna* L.), henbane (*Hyoscymus niger* L. + spp), and hemlock (*Conium maculatum* L.), the latter famous as the fatal poison drunk by Socrates. As benign and useful as a drug it may be still, plants containing pharmacologically active concentrations of atropine and related tropane alkaloids are, nevertheless, deadly poisons. The difference between a drug and a poison lies in the concentration, administration (applicant or oral), and frequency. As we examine mandrake's historical uses, do keep this fact in mind.

Even though the primary pharmacologically active constituents are atropine, hyoscyamine, and scopolamine, mandrake has more than eighty compounds. Whereas we may isolate a compound, even synthetically manufacture it, there is the possibility that other compounds act in synergy and affect the crude drug's action. To repeat: the tropane compounds act as a nonselective competitive antagonist of cholinesterase at muscarinic cholinergic receptors. By blocking muscarinic action, it inhibits sweat and saliva production and contributes to a relaxation of smooth muscles (known as parasympatholytic action), just as stated in the Assyrian recipes. Hyoscyamine bypasses the blood-brain membranes and affects the central nervous system.[68] Hyoscyamine sulfate is sold today as a prescription drug (Levsin is its proprietary name).[69] Atropine (and related compounds) is employed today in prescription pharmacy (proprietary name, AccuHist) but to be used with numerous precautions and special care.[70] Historically, it was used also as sedative (in small amounts), anesthetic, and anodyne. James Duke, American's premier scientist on medicinal herbs, has listed among the actions of the mandrake: anesthetic, cathartic, cholagogue, emetic, hypnotic, mydriatic, narcotic, poison, nervine, purgative, refrigerant, sedative, and stimulant.[71] To this list I shall add anodyne and soporific, perhaps imbedded in actions Duke says.[72] In low doses of mandrake's compounds, the effects are virtually limited to the stimulation of vagal (nerve) and respiratory system, but in higher dosages disorientation, hallucinations,[73] and comatose state ensues, and, still higher, death. Also, presumably on the basis of historical use, Duke said that mandrake was an aphrodisiac, a subject to which we shall return after examining historically one leading action, as an anesthesia, and general uses as given by prominent classical medical authorities.

"The Gift of Showing Mercy"

A modern anesthesiologist, Robert Holtzman, explored mandrake's use in surgery.[74] Its use in Western medicine continued until shortly after 1846, when volatile anesthetic drugs replaced it and other natural drug agents. John Scarborough, a distinguished historian of ancient pharmacy and medicine, has a far more detailed study on mandrake in ancient surgery.[75] Popular wisdom holds that before the late nineteenth century, deep surgery, especially abdominal and limb amputation procedures mostly entailed having the patient endure the procedure while a surgeon hurried the process quickly, not only for his patient's pain but also to diminish death by shock. A personal note on my family history: my great-grandfather who lived to within two years of my birth, received a gunshot wound at Chickamauga in the American Civil War. He lay on the battlefield for two days until he was found and carried to a field hospital. The surgeon took one look and said that the leg must be amputated; he pointed to the whiskey barrel to which my ancestor crawled and drank. The more he drank the more he insisted that the leg would not come off. The surgeon said that was fine with him, but he issued his prognostication: death. My grandfather did not die but lived with the lead bullet in his leg and took progressive amounts of opium until his death in a fire in 1935. From the nineteenth century through the present, atropine drugs were not abandoned; they were administered as a premedication drug to prevent side effects of anesthesia.[76] The point of the story is this: why, when atropine was available and better than alcohol, was it employed less frequently from the classical era to the Civil War?

Greek Medical Authorities on the Mandrake

Let us survey some leading classical medical authorities on the mandrake's actions before addressing the question about reproductive physiology. The general pharmaceutical usage underlying the discussion is as a fertility enhancer. We begin with "Hippocrates."

Hippocratic Works (*ca.* 440–330 BCE): The earliest surviving Greek medical writings of any consequence are those attributed to Hippocrates, the Father of Medicine, but these works were actually composed by many writers in the fifth and fourth centuries BCE. Mandrake was not a prominent drug, but it clearly had use as a strong narcotic and sedative. One having quartan fever (malaria) was to take a prescription to purge his bowels and then his head; if the fever persisted for two more days, he was to take henbane seed equal to the quantity of millet-seed, the same amount of mandrake [presumably the root], the mandrake juice in the amount of three beans, and the same quantity of clover juice. These drugs were mixed with undiluted wine and drunk.[77] Quartan fever is caused by mosquitoes carrying *Plasmodium malariae*,[78] a parasitic protozoon, and is also known as "benign malaria," since the infection, although long-lasting in seventy-two hour periods (hence, "quartan" or "four"), is seldom fatal. Fevers occur for day-day intervals, hence no fever on the fifth day. The laxative and headache remedy is mild, and the prescription containing mandrake (and henbane, which has a similar action) is in a small dosage and given only after the third or fourth day, when rest would be helpful for recovery but would not interfere with immune response to the infectious agent.

Mandrake's use as a drug in other Hippocratic works is consistent with its narcotic and sedative effects. Examples: take for the treatment of madness[79] with a tendency toward suicide the root of the mandrake in a small amount in the morning should the person become delirious.[80] For a condition of vaginal bleeding, mandrake juice was applied on wool as an applicant to enable the woman to sleep.[81] For an inflamed rectum fresh mandrake juice should be boiled in diluted wine and applied as a plaster.[82] For a person whose chronic illness makes him suicidal, he should drink mandrake juice in an amount less than what would lead to delirium. Similarly the juice should be given to one who has had a convulsion, and, again, in an amount less than that causing delirium.[83] A pessary was given to awaken or to stimulate (*energon*) a woman with mandrake juice and colocynth (*Citrullus colocynthis*) given in milk.[84] But, in the lengthy three-book treatise on women's diseases, mandrake is not included among the drugs that help women to conceive. Does this mean that mandrake was no longer employed to enhance fertility? No, it means only that the surviving Hippocratic works do not prescribe it for fertility; where the Hippocratic writers employed the drug, its use is rational according to our understanding of its actions.

Theophrastus (371–287 BCE): The author of the earliest herbal appearing as Book Nine of his *Enquiry into Plants* has scant mention of mandrake's medicinal qualities, unlike most herbs that have more detail, but he includes it in a list of four plants that induce sleep (*hypnos*) and act as a love potion (*philtron*). Theophrastus did, however, relate folk use about the mandrake with similarities to Harry Potter story that we saw earlier in this chapter.

Dioscorides (fl. 50–70 CE): As the author of a pharmacy guide in five books, Dioscorides was considered the foremost authority on the subject for 1,500 years. There are two kinds, male and female, he repeated, because the gender differentiation of mandrake plants was part of the folklore in antiquity. A chemical study of two distinct species (*M. autuminalis* Bertol. and *M. vernalis* Bertol.) showed some differentiation on the basis of their alkaloid content but within the range of normal biological variation.[85] In other words, the differences between "male" and "female," assuming the differences were botanical, would have been negligible. Dioscorides gave instructions about the preparation of mandrake juice as a drug taken with sweet wine and three *cyathoi* [=.approx. 0.137 liters] for those about to undergo surgery: "for they become unaware of the pain because they sink into deep sleep."[86] Just as did Theophrastus, Dioscorides related that the mandrake root was "good for making love-potions."[87] The organization for relating natural-product drugs devised by Dioscorides was to arrange by drug-affinities. Thus, if one did not have access to one drug, the drug either just in front or behind had similar physiological effects and might be a substitute. Before the mandrake (4. 75) Dioscorides described *Convolvulus ovalifolius* L. (4. 74) as a soporific (sleep inducing) and for "love-potions," but he cautioned, if taken in excess, the plant can be fatal. Aphrodisiacs are separate functions from aids to conception, although there is the obvious connection. Indeed, Dioscorides specifically recommends three drugs that "aid" conception: darnel (*Lolium temulentum* L.), wild carrot (*Daucus carota* L.), and rabbit rennet.[88] Rennet, Dioscorides says, "aids conception, but if drunk after menstruation, it causes barrenness."[89] Dioscorides made distinctions between an aphrodisiac (or love-potion or philter) and an aid for conception. Peter V. Taberner studied aphrodisiacs historically and concluded that there is no known mechanism of action,[90] although since the publication of his study (c. 1985), we have seen the drug, sildenafil citrate under the proprietary name of Viagra. Sildenafil citrate was first given for hypertension and angina pectoris, and, after usage, was found good for erectile dysfunction. The compound is a synthetic; no plant that I have found mimics its action. As Michael McCormick said, "The absence of evidence is not evidence of absence."[91] Among the ways aphrodisiacs may have "worked" is simply to make a person more receptive to sexual relations.

Leah's and Rachel's problems were not a deficiency of desire but an inability to conceive. Dioscorides distinguished between aphrodisiacs and aids to conception and so shall we. The remainder of Dioscorides's chapter on the mandrake is generally consistent with what we know of its medicinal actions: soporific, analgesic, anodyne, with ophthalmologic applications.

Galen (ca. 129–post-210 CE): Galen is the classical world's preeminent medical writer, based on voluminous surviving works in one edition of twenty-two volumes in modern print and still there were more titles omitted in that edition. Throughout his works there are infrequent mentions of mandrake and, when he observed it, generally it was to place it in the category of extremely cooling remedies and poisons. Following the humoral theory of addressing drug therapies by opposites, under narcotics (our term) Galen (and his sources) placed mandrake as cooling because of its dulling, soporific, and depressant to the nerves. Often he listed mandrake together with opium poppy, hemlock, and henbane and explained that, while useful as drugs, they were potentially fatal.[92] Similar to opium, mandrake may feel warming to the patient but its effect is extreme cooling.[93] It works as an anodyne,[94] but it leads to extreme constipation and a comatose state.[95] Notably, Galen prescribed its use neither to enhance fertility nor as an aphrodisiac.

Soranus (fl. 98–138 CE): Soranus is the greatest medical writer on gynecology to survive from antiquity. Soranus covers most aspects of women's medical and reproductive problems in detail (including contraception and abortion), but, on the matter of infertility, he is mostly silent, as was Galen. Women of childbearing age who are sterile, he briefly stated were undernourished, extremely thin, or very fat.[96] Using an earlier authority, Diocles of Carystos (fl. fourth century BCE), Soranus related a test for fertility: give to a woman a vaginal suppository consisting of such substances as resin, rue, garlic, nosesmart (a type of cress in the genus Nasturtium), and coriander. If the woman tastes the substances through an upward flow to her mouth, she is fertile.[97] Neither Galen nor Soranus had treatments for infertility. One explanation for the paucity of attention given to fertility enhancers in classical sources and the greater attention in those of ancient West Asia and Egypt is that the cultures gave little emphasis to childbearing and, indeed, the lower birthrate was a cause of concern for the Roman Imperial government. Dioscorides, on the other hand, named the following substances that cause barrenness (infertility): juniper (applicant to vagina;

1.77.2); white poplar (orally); rabbit rennet (2. 19; 2.75), menstrual blood (applicant; 2. 79. 2); cultivated cabbage (as pessary, 2. 120.3); pepper (applied after intercourse; 2. 159.3); ivy (drunk after cessation of menstruation, 2.179.3); green mint (pessary before intercourse, 3.34.2); axe weed (*Securigira securidaca* L; pessary before intercourse, 3. 130); miltwaste (*Ceterach officinarum* Willd. hung by itself or with a mule's hoof, 3. 134. 2); *Epimedion* (?, 4. 19); bracken (*Pteris aquiline* L., orally causes both barrenness and an abortion); iron rust (5. 80); alum (5.106. 6—also used for same in Assyrian recipes); and clay stone (5. 146, drunk for four days). Of these, eight substances were also either contraceptives or abortifacients. The amount and frequency would determine whether infertility was temporary or permanent. The point to be made is that Dioscorides believed infertility could result from drug substances purposefully or inadvertently taken. Several of the above Dioscorides related because of folk-belief, not that which he endorsed them, since he added such phrases as: "it is believed...." and "they say...." The remarkable point is that Dioscorides, Galen, and Soranus gave little attention to infertility reversals.

Latin Medical Writers

Pliny the Elder (23–79 CE) wrote *The Natural History,* a general encyclopedia of lore, in which he made many references to mandrake in treatments for scrofula, superficial abscesses (applicant), gout, erysipelas, and ulcers.[98] Under diseases of woman he listed mandrake juice orally to purge the uterus (emmenagogue), to remove a dead fetus, and to check excessive menstruation, but not to promote fertility.[99] He named two unidentified herbs: *arsenogonon*, which, when drunk, causes the conception of a male and *thelygonon* the conception of a female "if we care to believe it," he added.[100] But under wines he was probably referencing mandrake wine without naming it when he wrote: "A kind of wine in Arcadia [a region in Greece] said to produce ability to produce children in women and madness in men."[101] Reasonably, this wine may be assumed to be the famous mandrake wine that earlier we saw about mandrake in ancient Egypt. Similarly Celsus (*ca.* 25 BCE–*ca.* 50 CE), a Latin medical writer, picked up some folklore when he wrote that some induce sleep by draughts of poppy or hyoscyamus, while others put mandrake apples under their pillow for sleep.[102] Columella referred simply to "the maddening flowers of mandrake."[103] (The hyoscyamus plant has virtually the same medicinal effects as mandrake.) Just as with the Greek medical writers, the Latin writers were virtually unmindful of fertility enhancers.

Classical Anecdotal Data

Even more than the classical medical writings, anecdotal information sheds more light on mandrake use (including as a possible fertility enhancer). Socrates was discussing wine-drinking at a symposium when he said (according to Xenophon): "For wine does of a truth 'moisten the soul' and lull our griefs to sleep just as the mandragora [mandrake] does with men, at the same time awakening kindly feelings as oil quickens a flame."[104] Ironically, Socrates would die from a lethal dose of hemlock that contained many of the same compounds as those found in the plant, the mandrake, about which he had spoken so fondly. Another author, Apuleius, told a story about a villain seeking to remove one of his enemies by asking a physician (*medicus*) to supply him with a deadly poison. The doctor acting on his understanding of the Hippocratic Oath, said: "And so I [gave him] a drug, but a soporific, mandragora, well known for its proven lethargic effect, which produces a coma very like death."[105] The happy result for all but the villain was that the poisoned boy, the physician speculated, would merely go to into a comatose state for a long enough period for the villain to celebrate before the boy would awake from a long but pleasant sleep. Mandrake's strong soporific qualities are discussed in a number of ways by such authorities as Plato, Demosthenes, Aristotle, Philo Judaeus, Polyaenus, Lucian, Plutarch, Macrobius, Cyrillus (*Commentary on Genesis*), Theodoretus, and Frontinus.[106] Julius Caesar cleverly saved himself by using mandrake wine to capture his enemies. When he was taken by pirates as a youth and ransomed for too little, he thought, young Julius doubled the ransom and planned a farewell banquet for his captors. He served mandrake wine and, when they fell into a stupor, he had them captured and executed.[107] Mandrake's soporific use is well attested in classical literary sources including those as widely apart as Aristotle (fourth century BCE) and Isidore (seventh century CE).[108]

Frontinus (*ca.* 50–103 CE) told about how Maharbal, Hannibal's cavalry general, tricked the rebellious Africans around Carthage. He feigned a retreat and left in his camp copious amounts of mandrake-laced wine "which in potency is something between a poison and a soporific [*inter venenum ac soporem*]," Frontinus related. His enemies captured the camp, drank the wine, and fell into a stupor. Upon sobriety, they were prisoners.[109] Alexis (375—270 BCE), a New Comedy writer, wrote a play entitled "The Mandrake," wherein the heroine drugged herself for a plot that is too obscure to reconstruct.[110] While not naming mandrake, Caelius Aurelianus said that madness (*mania*) can come from "the drinking of drugs, especially those intended to excite love."[111] Plutarch, normally a rationalist, delivered the most unbelievable story: mandrake is so powerful that, when it grows near grape vines, it imparts its soporific forces through the soil and into the wine.[112] And Hesychius provided the shortest reference: "A kind of plant, intoxicating (? *oinikon*) and soporific (*hypnōtikon*)" and curiously said it was also called "The Zeus."[113]

MANDRAKE AS POISON

Some modern writers state that mandrake as a poison receives more modern attention than it deserves based on confusing gossip about the Roman imperial household. A number of respected Roman historians—Tacitus, Suetonius, Dio[114]—willfully related gossip by alleging that Livia poisoned Augustus and Agrippina poisoned Claudius; modern authors have refined the vague accusations to assert that these women used mushrooms and figs laced with poisons of mandrake and/or belladonna.[115] While the descriptions may have pointed to mandrake or mandrake-related poisons, the clandestine poisonings in the imperial household are probably fiction, good fiction, but fiction nevertheless. M. P. Charlesworth long ago, before Robert Graves spun the delightful novel *I, Claudius*, said the allegation is "utter absurdity and baselessness of such a notion...it is merely part and parcel of the ridiculous *canard*" that the emperors and their wives were notorious criminals.[116] Mandrake is more famous in modern writings about the imperial period than it deserves to be.

THE "APPLE" AS FERTILITY ENHANCER IN THE GARDEN OF EDEN

A collection of tales known as *Physiologus*, as we have it, was written in Greek and Armenian probably in the second (some say, third) century of our era and tells information, much fanciful, about animals whose traits exemplify traits that good people should emulate. A Hellenistic Jew may have been its first author, especially considering the biblical quotations and allusions, but the version that we have is clearly Christianized. If it existed in Hebrew, we have no copy. Whether Christian or not, it was readily adapted to Christian precept education and eventually elements incorporated in a genre of literature known as bestiaries.[117] The Greek and Armenian versions do not include the mandrake, but the Latin bestiaries, derived from *Physiologus,* tell about the elephant.[118] The elephant is drawn beside a male and female mandrake. The tale startlingly brings us back to the mandrake as a fertility enhancer and as the fruit of knowledge in the Genesis Garden of Eden story (Chapter 2).

> Now the Elephant and his wife represent Adam and Eve. For when they were pleasing to God, before their provocation in the flesh, they knew nothing about copulation nor had they knowledge of sin. When, however, the wife ate of the Tree of Knowledge, which is what Mandragora means, and gave of the fruits to her man, she was immediately made a wanderer and they had to clear out of Paradise on account of it. For, all the time that they were in Paradise, Adam did not know her. But then, the Scriptures say: "Adam went in to his wife and she conceived and bore Cain, upon the waters of tribulation."[119]

Another version has more detail:

> The elephant and his mate go there, and she picks a fruit from the tree and gives it to him. And she seduces him into eating it; after they have both eaten it, they mate and the female at once conceives...The elephant and his wife represent Adam and his wife, who pleased God in the flesh before their sin, and knew nothing of mating or of sin. When the woman ate of the tree, that is, gave the herb *mandragora* which brought understanding to her husband, she became pregnant and for that reason left paradise.[120]

Theobaldus (eleventh century) wrote a Latin version of *Physiologus* in which he probably collapsed the details, but Theobaldus said that Adam "fell because of the wood of a tree (*per lignum*)," not naming the tree but tying the disobedience to the tree of knowledge.[121] Michael Glykas, a Byzantine historian writing about 1118, wrote in reference to the mandrake: "Clearly this is a symbol of our first parents, for before these had tasted of the fruit of wood, the life they led was even as that of the angels, but when they tasted it and the transgression had been committed, then Adam knew Eve as his wife."[122] An earlier Latin version identifies the "tree" as the mandrake: *quando autem mulier manducauit de lingo (hoc est intellegibilem mandracoram)*.[123] Hildegard of Bingen (eleventh century) believed that the mandrake was the clay that God used to form Adam.[124] Max Wellmann believed that the mandrake-Garden of Eden story goes back to an unidentified Jewish source.[125] The Syriac version of *Hermes Trismegistos* contained a story in which King Solomon had a mandrake in his signet ring whose miraculous qualities enabled women to give safe birth without pain.[126] A later medieval German text related to *Physiologus* says that when a man wants children he takes his mate to paradise there to find a mandrake.[127]

The Latin bestiaries incorporate the elephant story. The bestiaries proved a very popular literary form from the eleventh century. Christian symbolism easily made elephants as the model for ideal human behavior because they were "devoid of all appetite for sexual intercourse" and, when they occasionally had sexual relations, the purpose was solely procreation[128]; they enacted procreation by going to paradise and eating the mandrake.[129] The elephant was "the Ideal Spouse."[130] Most bestiary versions repeat that the elephant represents Adam and Eve when they were "pleasing to God" and did not know about copulation nor could they sin. Eve ate of "the Tree of Knowledge, which is what Mandragora means" and immediately gave it to Adam. For this act (the text drops the serpent and disobedience), they were expelled from "Paradise." "Adam went in to his wife and she conceived and bore Cain, upon the waters of tribulation."[131] This is amazing: not only is the Forbidden Fruit the mandrake but the plant imparted "understanding" to both her and Adam and, with it, she became pregnant. And, according to another version quoted earlier, Eve's act of eating the mandrake caused her and Adam to conceive, but her persuasion of Adam to eat it in the first place caused him to have understanding (reason).[132]

Returning to the Question about Mandrake and Fertility

How does one reconcile the omission of mandrake as a fertility agent in the medical writings and literary anecdotal writings? We have come virtually full circle from the incantation of Inanna where she invokes, I postulate, the pomegranate and the mandrake, to the medieval stories that equated Genesis' tree of knowledge with the mandrake. Moreover, in an early *Physiologus* version, the mandrake enabled Eve to conceive. How do we reconcile mandrake's omission in the medical writings for fertility, when these same writings prescribe its use as a soporific, narcotic and in other ways compatible with modern science's findings? Both literary and medical sources have allusions to its use as a love-potion but there is nothing explicit about its enhancing fertility, as stated in Genesis.

Syriac Medicine

Probably in Edessa, fourth century CE, an unknown author or authors wrote in the Syriac language a fairly comprehensive "Book of Medicines." Throughout, there are prescriptions for mandrake in much the same way as with Greek medical writers, some of whom served as sources for the Syriac author(s). Appended to the manuscript were miscellaneous native prescriptions, among which was this one, the first herb of which cannot be easily identified:

> Concerning the root *sĕphînâytâ*. It is yellow, and is like hyoscyamus, and it beareth berries. It is good for the man who hath jaundice. Take of it a piece of the size of the chickpea, and throw it into wine, and give it to the patient to drink morning and evening, for two days. It is good for [the bites of] snakes. Take the seed of this root, and the seed of hyoscyamus, and the seed of *mandragora*, that is to say, of *tûryagh*, pound it, and drink it in wine, and thou wilt not wake up for three days. The root of this plant and its seed are good for inducing conception in women.[133]

Sĕphînâytâ's description of medicinal value from root and pale yellowish flowers fits the belladonna, whose medicinal action is much the same as mandrake (*mandragora,* whose flowers vary from pale white to light yellow). The Syriac source may have meant differing species of mandrake, as classical sources probably did when they referred to the male and female mandrakes. Even if the plant is not mandrake (since mandrake is specifically listed), the likelihood of its being belladonna further reinforces our interpretation. The mixture of belladonna, hyoscyamus (*Hyoscyamus spp.*), and mandrake is virtually mixing the same active compounds. Hyoscyamus is of the same Solanaceae family. Thus, after an absence of more than a thousand years in the surviving medical literature, we find the mandrake and plants with similar compounds as "inducing conception in women." Now, we return to the question that began the chapter: did mandrake actually work to enhance female fertility?

The Mystery of Mandrake's Fertility Action

Most modern writers on the mandrake, if they even speak to the subject, interject that there is no known pharmacological action for the mandrake causing an infertile female to conceive. Such was the muse of the eminent physician and scholar of biblical medicine, Samuel S. Kottek.[134] Julius Preuss, a scholar of biblical botany, opined that there is "no evidence for or against the efficacy of the mandrakes." He asserted, however, that no medicine can be effective without God's blessings but "even God employs natural means."[135] In speaking of Leah, Rachel, and the mandrake, St. Augustine of Hippo wrote: "Some people hold the opinion that this fruit appears to have the ability to make a sterile woman fertile... but I do not agree with this."[136] He said that he had examined the fruit by its sight, smell, and taste and found no quality that would explain the action. Instead, he said that God, not the plant, had opened the wombs. St. Augustine was arguably the Christian world's foremost theologian, but herbal medicine did not rank high in his learning. Going beyond—way beyond—Augustine's research methodology, I searched the library stacks, including those shelves with medical and pharmaceutical journals and the Internet (where mandrake lore is manifestly popular). In addition, I sent email queries to a number of fertility and herbal experts. No one cited a laboratory study. For all who answered—most did—I thank you without giving your name. All of the few scientists from whom I sought permission to cite agreed.

Mechanism for Action as Fertility Enhancer

Four medicinal chemists in the School of Pharmacy, Hebrew Union University, Jerusalem, conducted a thorough review (in 2005) of mandrake's chemistry. To be sure, these chemists gave attention to the biblical accounts that the mandrake promotes female fertility. They concluded: "The chemical principle of its 'magic' (hallucinogenic and poisonous) properties is today well known, but to understand its 'aphrodisiac' ('love') properties, still more analytical work would [*sic.* "needs to be"] done."[137] What little attention chemists have given to mandrake as an anti-fertility agent—and that is virtually none because all the guides say in effect its action, if any, is unknown—has been because of its action as an anticholinergic drug. I propose its action is more related to its attested "magic" actions, as these chemists called them.

Proposed Mechanism for Action

Writing for *WebMD*, Colette Bouchez shares these reactions given to women who were unable to conceive:

> You're just trying too hard.
> You're too stressed. Just relax and it will happen.
> You've got to calm down and let nature take its course.[138]

To this list, I add this one because I have heard it frequently:

> Just take out adoption papers and see if you don't get pregnant.

This same article says that approximately 30 percent of all infertility problems are attributable to stress. A statistical study conducted on the psychological impact of infertility found stress attributed to infertility is a factor that is statistically equal to stress felt by patients with cancer, hypertension, and heart disease.[139]

Infertility carries serious psychological impact in modern studies with contemporary patients, even though many infertile couples profess to have marital and sexual satisfaction. Ovulatory function is affected by stress and in no minor way.[140] A medical authority, Frederick Kakis, states flatly: "Female infertility is always in one way or another linked with problems of ovulation."[141] Stress causes or is a major part of an estimated 30 percent of infertility problems in modern women.[142] Whereas designing clinical trials for a human is a difficult way to test the working hypothesis, doctors of veterinary medicine are persuaded that the "stress factor" is "very real in animal infertility," according to Robyn Wilborn and Brian Whitlock of Auburn University. Even though stress is different in animals compared to that in humans, stress produces the same physiological effects (cortisol and endogenous opioids) that suppress gonadotropins, which make ovulation possible.[143] Stress activates the hypothalamic-pituitary-adrenal (HPA) axis on human female reproductive systems. In the technical terms of a recent journal article on humans: "Corticotropin-releasing hormone (CRH) inhibits hypothalamic gonadotropin-releasing hormone

(GnRH) secretion, and blucocorticoids inhibit pituitary luteinizing hormone and ovarian estrogen and progesterone secretions."[144] The pituitary LH (luteinizing hormone) is necessary for ovulation and ovarian progesterone and therefore necessary for maintenance of pregnancy.

Conclusion

Stress affects modern women's fertility, say many modern studies. If as many as 30 percent of modern women are infertile because of stress as a major or, in some cases, exclusive factor, one needs a multiplier to measure the stress on the wives of Abraham, Isaac, and Jacob. Not only was fertility of paramount importance in their society: God had commanded them to multiply and, in addition, God prophesied that the children of their husbands would form a nation of chosen people. I cannot think of a more stressful situation for those women. A tranquilizer would definitely be in order—or anything else to reduce stress![145] The classical medical sources eloquently testify that a small amount of mandrake was useful to relax and to induce sleep—what we call stress therapy. The mandrake that Reuben found in the field was just such a tranquilizer for Leah and Rachel and presumably for other women in the former matriarchal society as it turned strongly patriarchal. The ancients used mandrake as a drug in most ways that we know to be rational. Why should we not trust the Bible when it says the plant gave to the Jewish nation Joseph and four of its tribes? From this lineage came the House of David and the family of Joseph, Mary, and Jesus. God opened Sarah's womb (possibly naturally through the mandrake) and from her womb came the lineage for the Jews, Muslims, and Christians. So it is believed. Like the pomegranate, the mandrake plant was an important factor in human history. Mandrake was a powerful drug but it was misused and came to be used by the devil; so it was believed in the Middle Ages, and, as we shall see in the last chapter, there were grounds for that belief also.

Epilogue

A few years ago a group of American students and I visited the Egyptian Coptic monastery of St. Mina (or Menas) in the Amriya district in the desert roughly between Cairo and Alexandria. The monastery is a famous pilgrimage site for infertility. As pilgrims, infertile people will be given a vial of St. Mina oil. Accompanying us was an Egyptian Coptic Christian who had his son, age seven (as I recall), who was a product of St. Mina's miracle. The couple was infertile; incidentally our friend was trained in medicine. He and his wife visited the monastery and received the holy oil. Shortly thereafter a son was conceived whom they named Mina after the Saint. Monks at the monastery told the story of a man who, in a similar condition, received the oil but returned approximately ten years later, reported that he and his wife had had seven children, and he now wanted an oil to reverse fertility. The monks told him the miracle worked in only one direction and could not be reversed.

Why tell this story in the context of mandrake? I am aware of the positive effects that placebos can have and how important religious devotion and prayer may be. That acknowledged, one placebo is as good as another (provided no harm, of course). The mandrake was not a placebo; it had positive pharmacological action that is based on the weight of historical evidence and on the understanding of the modern science of reproductive physiology.[146] The mandrake assisted in conception when stress was an inhibiting factor.

While we reserve the final chapter for reporting how mandrake came to be perceived as the Devil's helper, not the positive joy of children, the mandrake survived the medieval period as a female fertility agent. Frustrated by his inability to persuade contemporary Florentines on political issues (as he did in the *Prince* and *Discourses on Livy*), Niccolò Machiavelli (d. 1527) turned to comedy as a trial vehicle. *Mandragola* ("The Mandrake") is a play in which he said in the Prologue: "*La favola Mandragola si chiama: la cagion voi vidrete*/ Our tale's named for the mandrake root. You'll see the reason as we play it."[147] The plot turned on the mandrake's fertility enhancing qualities. Two clever guys sought to seduce a beautiful, faithful wife of a wealthy merchant who was childless. The clever ones persuaded the husband to give his wife a potion made of the mandrake root for her to conceive—the husband believing that he was the most virile man in Florence and therefore could not be the problem. The hitch was that he is told that, while it will cause his wife to conceive, the first person to have intercourse with her following the mandrake's administration would die within a week. Thereafter, he could have safe sex with his wife and conceive. Well, not to betray the play's unfolding of events readily seen by modern sophisticated readers, Machiavelli knew already that his audience would know the mandrake's effects because he did not elaborate with explanations to enlighten his early sixteenth-century audience. Using poetic license, he modified its death-inducing legend in order to have his character seduce (with happy results) a beautiful woman, showing, I presume, what a clever prince could do with policies by "spinning" parts of "truth" in order to obtain desirable objectives.

"No lustful woman came this plant to grieve," John Donne (d. 1631) explained.[148] In closing, here is my favorite mandrake poem by Donne who knew and believed in the mandrake's fecundity: His poem:

> Go and catch a falling star
> Get with child a mandrake root
> Tell me where all past years are,
> who cleft the Devil's foot.[149]

CHAPTER 4

ARTEMISIA, THE "MOTHER HERB"

> There are some who think that the surname [for the plant] is derived from Artemis Ilithyia, because the plant is specific for the troubles of women.
>
> —Pliny, *Natural History,* 25. 34. 73

Ilithyia, or Eileithyia, was the goddess of childbirth and labor pains, and, as Pliny implied, the deities Ilithyia and Artemis eventually became one. When they were separate goddesses, Artemis was the more honored. St. Paul encountered Artemis worshippers in Ephesus and spoke of the intensity of their devotion (Acts 19: 23–40); in assembly for two hours the Ephesians chanted: "Great is Artemis of the Ephesians!" "She whom all Asia and the world worship," Paul was told as he valiantly attempted to convey another message. Some authorities say that Artemis was the most popular of the Greek deities whereas others cautiously say one of the most popular. In her earlier form, she was an agricultural goddess in Arcadia (Greece) where she also became the goddess of the forest and hunt. Some ancient accounts present her as the daughter of Zeus and Demeter, or Persephone (who may be conflated into one). Her twin brother was Apollo beside whom she was frequently represented (as we shall see below on the Selinus coin). Before marriage women gave clothes to Artemis, and in turn, she protected pregnancies. After giving birth, women dedicated their prenatal clothing to the goddess. The temple functioned as an ancient equivalent of Goodwill Industries. Women and men associated Artemis with the woodlands and her being surrounded by nymphs. Artemis was "deadly and remorseless toward those who threaten her chastity" and, with equal energy, she protected the young and mothers of all animals, even though, seemingly incongruously, she headed the hunt.[1] A later legend about a woman named Artemisia, who married Mausolus (d. 353 BCE), dynasty of Caria, was later confused as the person who gave the plant its name. In legend or in fact, Artemisia was so grieved by the death of her husband that she gave him a magnificent tomb, hence the name "mausoleum."[2] This story confused many later writers, such as Gerard in his 1636 Herball who wrote: "Mugwort [A. vulgaris L.] is called in Latine, Artemisia, which name it had of Artemisia, Queene of Halicarnassus, and wife of noble Mausolus, King of Caria, who adopted it for her owne herbe."[3] However, in antiquity, the name for the herb came from the virgin goddess, not the queen and wife of Mausolus.

Artemis could be vengeful; for example, she ordered Iphigenia's sacrifice before the Greeks could sail to Troy in Euripides' play Iphigenia at Aulis. Later, after Euripides and as a result of the ancient equivalent of political correctness, Artemis was said to have taken pity on the young woman and substituted a deer as a sacrifice. When Leto gave birth to Artemis, Leto suffered no pain; when Leto gave birth to Apollo, Artemis gave her mother assistance with a difficult birth.[4] Physicians, midwives, and "wise women" who assisted others in birth employed the plant for difficult births. Primarily, we surmise, this was why artemisia was called the mother herb. Always a virgin and subject to no man, Artemis's purpose was to protect women and wildlife.

In Sparta there was the festival of Artemis Orthia, when young Spartan boys would be seized and whipped, some to death. The origins of this brutal ritual was said to have been when a wooden statue of Artemis was kept in a grove of chaste trees (the subject for the next chapter). The flogging's purpose was purification of the Artemis altar in the tree grove by sacrificial blood. The ritual (known as the diamastigōis) was to keep the statue upright (Greek orthia meaning "upright"), [5] but whatever purposes the ritual began, it had different meanings as time progressed into the historic period. Inasmuch as Artemis was associated with childbirth and the perpetual virgin, I suspect that in time there was a perceived divine wish that young males should appreciate the pain of pregnancies and childbearing. As all too frequently happens, in time the religious purpose faded; by the third century CE a small amphitheater was built for the comfort of those who wanted to witness the sadistic ritual where boys met a cruel death by savage blows.[6] In Athens young girls (prepubescent and earlier adolescent) were sent to Artemis's shrine at Brauron, where they learned how to be spouses and working women in their society. Doubtlessly they learned about why the goddess plant would be important in their lives.[7] Beyond Athens, many cults for Artemis existed for women in transition, such as puberty, sexuality, birth, motherhood, and death.[8]

This chapter explores plants of the artemisia family in classical and, to some extent, Western medieval medicine for its total medicinal qualities that led to its name from the goddess Artemis. Chinese and Vedic (Indian) medicine will also be included. The herb was called the "mother herb"; like mothers, most of its various roles benefited human societies. In honesty, there were the hallucinatory and narcotic roles that harmed people. Several or more species in the artemisia family provided special comfort and assistance to women, ranging from abortion to helpful relief in safe childbirth. To men and women alike, its medicinal qualities range from a hallucinatory and narcotic effect to the treatment of intestinal worms, wounds, and fever—and this list will be expanded. From history's beginning to the present, artemisia's story is astonishing because it comforted and saved multitudes of women. To men and women alike, artemisia plants provided medical assistance in a variety of ways and, even now in the twenty-first century the plant saves the lives of hundreds of thousands annually from the wretched disease of malaria. Also, in fairness, its story will conclude this chapter with some adverse qualities. The plant is so hardy that the United States Department of Agriculture lists artemisia as "an obnoxious weed."[9] The plant is a miraculous drug in ancient Athens, but it is obnoxious to farmers in Alabama, or so goes its history. Its story is largely untold and unrecognized. Now in the twenty-first century, our attention to its qualities is returning in a vital way. Artemisia's modern use focuses on its malarial cure; its use in gynecology and obstetrics accounted for its greatest impact, although now largely forgotten. Between antiquity and today, artemisia had adverse qualities as a narcotic-like drug, which surely did as much harm as good.

ARTEMISIA PLANTS

Approximately four hundred species of the broadly distributed genus *Artemisia* are identified and more than sixty are used in traditional medicine, but the active chemicals are largely unexplored.[10] Before modern botanical nomenclature, people identified plants by common names. Certainly ancient people could not recognize or distinguish among all these species—nor generally did they need to do so. In this chapter we shall concentrate on those plants that we have thought were known, while, at the same time acknowledging that other, less prevalent species were subsumed under one name or another. In the classical period, three major names for artemisia species were used:

1. absinthe (Latin: *absinthium*) for our *Artemisia absinthium* L. Under this name there were probably these species that may have been at times subsumed under the absinthe name:
   a. Absinthium Gallicum and/or a. marinum, Gallic or marine artemisia (*A. maritima* L./ *A. Gallicum*, Marc.)
   b. Absinthium pontica or Roman wormwood (*A. pontica* L.)
   c. *Artemisia judaica* L./ alt. *A. herba-alba* Assoc. or Jewish wormwood/ white wormwood
2. southernwood (Latin: *abrotanum*) for our *A. abrotanum* L.
3. mugwort (*A. vulgaris* L.)
   a. Artemisia leptophyllos or our field wormwood (*A. campestris* L.)
   b. *Artemisia arborescens* (L.) or forest mugwort

Quite probably many ancient peoples did not distinguish southernwood from mugwort; therefore in this chapter we shall not attempt in most references to interpret which plant the ancient recorder had in mind inasmuch as the chemistry and medicinal uses were much the same. In contrast, absinthe was stronger and had pronounced hallucinatory effects but, generally, the ancients referred correctly to it and separated it from their kinds of artemisia with descriptive adjectives, such as Gallic, Pontic, and Jewish. Tarragon (*Artemisia dracunculus* L.) is today a member of the genus *Artemisia*, but the ancients considered it a different plant with different uses, often culinary; tarragon will not be included in this chapter. Also there is another species of artemisia (*A. annua* L.) with English common names of annual wormwood and Sweet Annie. This species is indigenous to Asia, especially Chinese and Southeast Asia areas, but it grows now in the Balkans, but we are uncertain whether it grew there in the classical period. Annual artemisia contains a compound in higher concentrations that is highly effective for malaria treatments. We shall postpone into later in this chapter whether annual artemisia was available and used by ancient peoples in Mediterranean-European climates.

Artemisia has distinctive characteristics, usually in a small bush, and its leaves are bitter; the range of habitats is virtually universal from deserts to the subarctic. Xenophon (d. 354 BCE), a Greek who visited Asia on a military expedition, said of Mesopotamia (he called it Arabia): "In this region the ground was an unbroken plain, as level as the sea and full of wormwood."[11] Xenophon used the word *apsinthon* (usually Romanized as *absinthium* and Anglicized as absinthe), loosely translated "wormwood," the English generic for most plants in the species. It is called wormwood because it prevents insects in clothes chest, insects when rubbed on the body in oil, and worms in the intestines. Let us start with the biblical bitter species, white wormwood or Jewish wormwood.

Jewish Wormwood and Artemisia Varieties in Ancient West Asia

The Bible has numerous references to absinthe as a synonym for bitterness as, for example, *Proverbs* 5: 4: "but in the end she [i.e., a loose woman] is bitter as wormwood (*laanah*), sharp as a two-edged sword." Also, as a mind-altering alcoholic drink spoken of in *Lamentations* 3: 15 and 19: "He has filled me with bitterness, he has sated me with wormwood"; "Remember my affliction and my bitterness, the wormwood and the gall." The biblical wormwood has been identified as *Artemisia judaica* L./ alt. *A. herba-alba* Assoc., or white wormwood.[12] The Hebrew word *laanah* is designated by biblical scholars as white wormwood, but the Septuagint and Vulgate use the Greek *absinthon,* which is another species, one much more potent in mind-altering qualities. All species have anthelmintic qualities, that is to say, it removed tapeworms and other intestinal parasites—afflictions that the ancients had plentifully and painfully.[13] These and related species were frequently given for a general term such as aids the stomach. Artemisia or wormwood is currently found effective for a number of digestive parasites, including trichinellosis.[14] The biblical references (also *Deut.* 29:18; *Job* 30:4; *Jeremiah* 9: 15, 23:15; *Hosea* 10:4; *Amos* 5:7, 6:12; *Revelation* 8:10–11) to wormwood bear out what we know about the plant.[15]

Artemisia is found in Assyrian prescriptions. The Assyrian word is *šam sîhu*, probably *A. judaica* and/or *A. absinthium* L., is found in Assyrian prescriptions for the feet, swelling, bruises, eyes, ears, coughs, and possibly pleurisy, apparently all used as applications.[16] The Assyrian prescriptions mirror those in the ancient Egyptian medical documents. The famous Ebers Papyrus (Recipe 56) lists the plants that expelled worms in the belly: 5 *ro* (an Egyptian measure; precise amount unknown) leaves of pondweed, 5 *ro* of wormwood, mixed and eaten. There is another for the same purpose: Recipe 64: melilot 1; wormwood 1; fermented plant juice 1: mixed and eaten. The Ebers Papyrus has at least four recipes using wormwood for internal parasites ("worms").[17] Similarly, there is the Assyrian recipe (*cf.* Ebers Recipe 138): for pain of demonic origin in the anus, wormwood 1/8th, juniper berries 1/16th, honey 1/32nd, sweet beer 10 *ro*; it is strained and drunk for four days.[18] The Ebers Papyrus recipes prescribed wormwood for general health, for the mouth, in various applications for wounds and abrasions, digestive problems, headaches, stopping bleeding (hemostatic), for toothpaste ("abrasive powder"), for "stopping the hardening in the limbs" (probably arthritis?), and for several nontranslatable diseases.[19]

For each of the ancient Assyrian and Egyptian medicinal applications for artemisia, there is adequate substantiation for its actions, possibly excepting toothpaste and the presumed arthritis remedy. Modern science studies are well documented and verify virtually all of the artemisia species' anthelmintic actions (against intestinal parasites),[20] with special attention paid by veterinary physicians because of artemisia's relative inexpensiveness.[21] As an application for the feet, swellings, bruises, and hemorrhoids (assuming that is "pain in the ass"?), compounds in most artemisia species are anti-inflammatory, antibacterial, antiviral, anticoagulant, and hemostatic.[22] The medicinal actions of southernwood (Latin: *abrotanum*; *A. abrotanum* L.) are mostly attributed to the choleretic activity of coumarin compounds found in this species: isofraxidin, scopoletin, and umbelliforone.[23] The plant's use for bronchial conditions is attributed to flavonol compounds found in this (and other) species: casticin, centaureidin, and quercentin derivatives, which have spasmolytic activity.[24]

Why the Mother Herb?

There are no surviving and translated records—thousands of tablets are untranslated—from ancient west Asia or Egypt where women were using artemisia plants specifically for female health. We know that the plant was named after the goddess and said to be the mother herb. Graeco-Roman records, however, make a strong case for why the goddess' plant was called the mother herb. Dioscorides, the herbalist of the first century CE, fails to tell us about a connection with Artemis; he said that it was used by the priests of Isis, the major Egyptian goddess: "use it instead of olive branches."[25] I surmise that the reference was intended as ritual use. Dioscorides visited Egypt and wrote that wormwood (*apsinthoun*) grew abundantly at Taposiris (modern Abusir near Alexandria) where the tomb of Osiris, Isis' brother was. The Ebers Papyrus connects wormwood with Isis in a prescription that Isis gave to Ra to cure a headache.[26] Artemisia plants, most especially *A. absinthium* L., were given for headaches for millennia and, as many people will attest, it also gave headaches or worse, if taken to excess.

Most classical medical writers do not state explicitly why the herb is special to women and to Artemis. Pliny said that its qualities were "specific to women."[27] Pseudo-Apuleius' herbal, written in Latin in late Roman-Christian period about the fifth century, under "Herba Artemisia Leptofillos [probably *A. campestris* L.]: Diana [Roman equivalent for Artemis] is said to have found three types of artemisia, the virtues of which, she gave to Chiron the Centaur who initially instituted their medical uses. For this reason the herb derives from Diana."[28] A ninth-century manuscript of this text has an illumination that shows Chiron holding three types of artemisia (see book cover). Chiron was the centaur who was the master of the healing arts but, alas, he could not cure himself. A late Roman writer of the fifth century said that the Greeks called the artemisia plant Chylon, presumably a corruption of Chiron.[29]

Artemisia's medicinal effects were what made the artemisia plants special to women. Dioscorides enumerated some of these qualities in wormwood while discussing *A. campestris* L. and *A. arborescens* L.:

> When boiled, they (two types) are suitable to use in sitz baths for drawing the menstrual period, afterbirths, embryos/fetuses, for uterine closing and uterine inflammation, for breaking stones, and for retention of urine. The herb, when liberally plastered against the lower part of the abdomen, sets the menstrual period in motion. Its juice, triturated with myrrh and applied, draws from the uterus as many things as the sitz bath; the foliage is also given to drink in the amount of three *drachmai* [a measure] to draw out the same.[30]

In separate chapters Dioscorides attributes the same menstrual problem solutions to southernwood (*A. abrotanon* L.) and absinthe (*A. absinthium* L.). Artemisia's anti-inflammatory, antibacterial, antiviral, anticoagulant, and hemostatic qualities doubtlessly contribute to its use for menstrual problems as the ancients attested. Today medicine containing or made exclusively of artemisia species are sold for premenstrual syndrome (PMS), as one can readily see by Googling artemisia and PMS.[31] One Web site simply describes artemisia as "A mildly tonic, gently detoxifying herb"[32]; another Web site proprietor sells white artemisia (same species as in the Bible?) as a "female cycle regulator."[33] Inexplicably, a leading, new, and otherwise good guide to healing herbs for women omits artemisia species from the list of herbs helpful to women.[34] Once, Artemis's herbs were women's paths to good health.

Women whose menstrual period was delayed, however, took the plant even when or, sometimes, especially when they perceived that pregnancy might be the reason for the cycle disruption. The plant in its various species was given from antiquity to the modern period to cause an abortion.[35] Studies show that it has 100 percent effectiveness in terminating pregnancies in rats, and another study shows that artemisia interferes with spermatogenesis. Generally a symbiosis exists between anthelmintic and abortifacient qualities: in other words, a drug that drives out intestinal parasites often terminates early pregnancies. The modern electronic guide to drugs, DrugDigest, specifies as the first precaution for artemisia-derived drugs: "Pregnant women should not take or use wormwood due to the risk of miscarriage. In animal studies, wormwood caused the muscles of the uterus to tighten, which could result in a miscarriage."[36] If we judge by the infrequency of recordings in medical and literary sources, artemisia was not a primary abortifacient; its major use: was for more ordinary use to help with PMS and obstetrical purposes. The sitz bath indicates rational use of its anti-inflammatory and antimicrobial qualities.

Soranus on Artemisia for Women

Soranus, the major authority on gynecology, prescribed absinthe in metasyncritic treatments (restoration of the pores and interstices of the body to a normal state), and for chronic menstrual problems (amenorrhea and dysmenorrhea).[37] A woman intent on aborting her fetus should have sitz baths two or three days in which there is wormwood (*A. abrotanum* L.; or *A. vulgaris* L) and absinthe, with the latter apparently being the stronger because it was used when milder wormwood recipes failed.[38] Soranus said that some physicians recommended sweet Cretan wine with southernwood (abrotanum) or absinthe to treat pica (a perversion of appetite associated with pregnancy) but he opposed the procedure because absinthe, at least, incurred a danger of a miscarriage.[39] Soranus did not exclude southernwood artemisia, however, from pica therapy. Inasmuch, as (we shall see below) absinthe has a higher concentration of a toxic compound, Soranus was very discerning.

Hippocratic Writings and Artemisia

Soranus disagreed with the Hippocratic writers who recommended southernwood to accelerate a slow labor; in *On Women's Diseases,* a surviving work ascribed to Hippocrates, there are prescriptions to accelerate birth, but neither southernwood nor another artemisia species is among the ingredients.[40] The same Hippocratic treatise, however, prescribed absinthe as a pessary and as a fumigant for difficult births (breech), a remedy to which modern science does not presently attest.[41] In the Hippocratic work *On Female Sterility,* absinthe was used as a fumigant and/or application for a woman's womb that resisted conception, to be administered for periods as long as four months.[42] In other Hippocratic works, the medical writers prescribed southernwood (*abrototon*) with pepper and honey in a lozenge for pneumonia and for pain resulting from pleurisy. Also, non-gender-specific applications were absinthe as a topical for tetanus and absinthe in a warm bath for chronic, painful jaundice.[43]

Some Other Classical Medical Writers on Artemisia

Galen was the great physician working primarily in the second half of the second century of our era who used southernwood and absinthe in much the same way as in earlier medical accounts: internal parasites, stomach, fevers, expels internal poisons through diuretic action, imparts bitterness in foods for "all hardnesses," and as an antidote against mushroom poisoning (both southernwood and absinthe).[44] He noted that the drugs from either plant are potentially dangerous and, even more so, is absinthe. The latter, he said, caused one to fall into a coma. (A later Latin version translated the Greek to mean "highly hallucinatory."[45]) Shortly we shall examine absinthe's hallucinatory/coma effects, but, for now to summarize, Galen did not have specific gynecological uses.

Celsus employed both southernwood and absinthe widely in medical prescriptions: southernwood for cleaning wounds and the body in general and for gout, absinthe for the stomach in general and for those filled with phlegm, increasing urine flow, jaundice, neck (when breathing is hindered), inflammations (in drink on empty stomach), liver disease, for spleen and intestines, for throat as a tonic, difficulty with micturition, intestinal worms, and as an external plaster for disease of viscera.[46] Curiously, Celsus often said to take absinthe on an empty stomach but modern people report that it is bitter and distasteful. As the old saying goes, if it tastes that bad, it must be a good medicine. Importantly for the section that follows on malaria, Celsus recommended southernwood in a pill for chronic diseases that supervenes on quartan fever, the latter being a form of malaria.[47]

Although not a physician, Pliny, always the dilettante, dropped considerable information about the use of artemisia. He employed two words, *abrotanum* (southernwood) and *absinthium* (absinthe) and named what we would call subspecies, such as Pontic wormwood (*Absinthium pontica* L.). Pontic wormwood, he said, provided a medicinal wine if one mixed one pound of wormwood with five gallons of wine. Southernwood was given for kidney problems, strangury (a urinary affliction), jaundice, emmenagogue (i.e., an abortifacient), "good for the uterus," the bowels and cleaning the ears, thus fairly well repeating previous accounts.[48] Perhaps the unique use was as an application to prevent chafing of the skin between the thighs. Absinthe keeps away caterpillars, abates a toothache, and is good for the liver, kidneys, dropsy, and excessive bile.[49] Yet, given all of these applications, Pliny said that artemisia plants were specifically for women's uses.[50]

Other medical writers used artemisia for breathing problems, hemorrhoids, worms, and, in wine, as an antidote for poisons.[51] Again, the applications exclusively for women were not prominent in most classical authorities, but why? The writers were mostly males, and women's problems were more the province of midwives. Dioscorides, Soranus, and Hippocratic writer of the treatise *On women's diseases* were the principal exceptions.

Artemisia's Use in Medieval and Early Modern Periods

Throughout the medieval period, both absinthe and southernwood provided prized drugs for essentially the same remedies as were in the classical sources. Remedies for gynecological problems were prominent in most works; usually the expression was "*provocat*/ stimulates" menstruation. Within the lexical range of this expression was abortion. Also, subtly, these herbs were said to "remove a dead fetus," a common circumlocution for abortion. Macer wrote a very popular herbal in Latin verse (eleventh century). His number one herb was "*artemisia*" (by which he meant absinthe), followed by "*abrotanum*" or southernwood. Absinthe was "the mother herb/ *herbarum matrum*." For both plants, his verses emphasized applications for women. The "artemisia" plant, Macer wrote, was named after the Greek goddess Artemis.[52] Macer's identity is not known but I believe that Macer's real name was Marbode, Bishop of Rennes, who wrote a similar verse on stones and minerals. Macer or Marbode purposefully invoked a pagan goddesses' name in an endearing way, a subject that earlier Christian sources avoided.

Al-Kindī (d. *ca.* 870), writing in Arabic, extolled southernwood's uses for the mouth (toothpaste, remove decay, halitosis, and general prophylactic),[53] and Maimonides, the great Jewish physician, repeated Dioscorides's recommendation for southernwood's astringent qualities but warned that it was bad for the stomach.[54] Ibn Sīnā (d. 1037; Avicenna, in Latin), the great Muslim physician, said that there are many kinds of artemisia plants, each with numerous mostly shared, medicinal virtues. Among them were stimulation of appetites, drunk before wine to prevent inebriation, various ophthalmologic uses, cures or prevents intestinal worms, helps digestion, and good for a number of special, female needs, such as stimulation and regulator of menstruation, and as a killer of harmful insects.[55] In Indian Vedic medicine, several artemisia plants (mugwort and absinthe among them) were similarly employed for intestinal parasites and as a tonic.[56] There are a number of Vedic works, however, where the drugs' identifications are uncertain in modern scientific nomenclature.[57]

Constantine the African (d. 1085) adapted a work in Arabic to his free Latin translation where he repeated many of the same things about absinthe as Ibn Sīnā; Constantine placed more emphasis on its gynecological applications, and he repeated the claim that absinthe prevented inebriation from alcohol. "Magister Salernus," an anonymous Latin source from Salerno, added that absinthe could be mixed with honey and placed in the vulva for stimulating menstruation.[58] In naming the remedies provided by artemisia plants, medieval medical works usually included the same as those in the classical works but perhaps with more attention given to their alleged properties to detoxify ingested poisons. A Salernitan work called *Tabula* says that absinthe "comforts the head and clarifies vision."[59]

Writing in Latin principally for the women in her convent, Hildegard of Bingen (d. 1179) called the plant *Wermuda,* the German word for absinthe and/or wormwood; Hildegard said that *Wermuda* was a principal women's remedy for all illnesses (*est principalis magistra ad omnes languores*). She related a long list of remedies that are much the same as those in classical sources. Hildegard received much of her information orally from her culture, not through written works; it is a reasonable conclusion that her information about absinthe came from general practices in medieval Germany rather than being derived from written sources. She added, however, that absinthe is given for mental depression (*melancholia*), a condition from which she suffered. She noted, however, that absinthe can be deadly.[60]

Platearius (d. *ca.*1161), the presumed author of an herbal called *Circa instans*, designated medical remedies for absinthe primarily for women. He began by asserting that absinthe acted against sterility that arose from excessive humidity (a humoral designation). The plant, he claimed, helped the complexion of women; a decoction of it with wine and laurel leaves was good placed on the vulva and also, in a different formula, could act in a suppository. Along with the usual remedies, Platearius recommended absinthe for *thenesmon*, a recent word for some kind of intestinal disorder.[61]

By the late medieval period, the plant (absinthe, wormwood, and southernwood) was given for chest congestion, coughs, lung diseases, and body pains, for much the same reasons as in earlier medicine, except it was usually made into syrup with other ingredients.[62] Nicholas Culpeper's *Pharmacopoeia Londinensi* (1649) said that southernwood combated poison, killed worms and insects—much the same as in classical sources—but added "provokes Lust." Similarly to absinthe, Culpeper added to the list of remedies: "cleans sight."[63] Culpeper was writing to combat the monopoly that apothecaries were increasingly making compound medicines by proprietary secret formulas. His counter-current movement demonstrates the continued importance of artemisia plants to which his age added new uses. The movement toward polypharmacy (using many simple medicines in complex formulae) was partly the consequence of the apothecary guilds developing secret proprietary mixtures to boost their economic interests as purveyors of drugs rather than the family herb gardens. He added treatments using artemisia plants, such as one for epileptic episodes. Hager's famous guide to drugs in 1874 said that artemisia plants were used for opening pores (*diaphoreticum*), against epileptic episodes, and as a promoter of memory.[64] Astoundingly (to me, at least) the conventional wisdom among the medical authorities of the ancient and medieval periods, absinthe and other wormwoods prevented inebriation and epileptical episodes, whereas (as we shall see below) in the nineteenth century these same plants were put in alcohol to enhance the initial psychological uplifting. When a person was drunk from the mixture (absinthe and distilled alcohol), they were said to have epileptic fits. Southernwood was used in obstetrics as well as the usual remedies such as anthelmintic and antispasmodic applications.[65] Especially using absinthe, a new means of drug delivery was developed in the late medieval period, namely distillation and combining with alcohol in various forms. A discussion about alcohol and absinthe will conclude this chapter.

The Christianized Plant of the Goddess in Ritual and Religion

Hildegard did not associate artemisia plants with the goddess Artemis, as Christianization of the culture had filtered out that knowledge except for a few scholars. Nevertheless, a name for mugwort (German: *Beyfuss*) was named after St. John and called *St. Johannsgürte* that means "St. John's girdle." Similarly in medieval French the plant was called "St. Jean." On St. John's day, the common folk tied bands of artemisia plants about their waist, hence "girdle." Wearing this girdle, they performed undisclosed rituals and spells, made rhymes, and ended their rituals by flinging the girdle into a fire. The ritual removed their hardships.[66] (Speculatively the girdle may have covered the genitals more than the waist.) A somewhat similar ritual was celebrated in Anglo-Saxon England, where the plant's name was *mucgwyrt* ("mugwort"). Walafrid Strabo (d. 849) composed a poem on the herbs in his monastic garden and spoke of absinthe's "marvelous" actions when worn as a garland (see poem below) for headaches. It seems likely that he was referring to an earlier version of the ritualized use of wearing wormwood girdles.

A young woman in Scotland was near death with pneumonia when a barmaid said to her providers a little verse that was repeated:

> Wad ye let the bonnie may die I' your hand,
> And the *mugwort* flowering in the land![67]

A Slavic term for the plant is *bozhye drutze*, meaning "God's little tree."[68] And in the Germanic Gothic of Sweden when one administers mugwort, called *grensink*, one should invoke the "Son Jesus Christ" or "the Father and the Son."[69] If one carries a twig of vitex (the plant which is the subject of chapter five) or artemisia, one's thighs will not chafe.[70] In a perversion of Pliny's recipe for artemisia preventing chafing of the thighs when applied in a salve, this medieval procedure of merely carrying it for this purpose is more magic than medicine. Whereas medieval culture mixed magic or superstition with medical practices (as did classical medicine in folk practices), artemisia plants were more connected to religion and ritual than was practically any other herb, doubtlessly as a vague memory that it is Artemis's herb. Artemisia plants were well known throughout the Middle Ages and through the Renaissance. Great botanical writers, such as Hieronymus Bock (d. 1554), Otto Brunfels (d. 1534), and Leonhart Fuchs (d. 1566) eulogized the "mother of all herbs."[71]

MALARIA TREATMENTS IN RECENT YEARS

In February of 2008, the *New York Times* (A4) had this headline:

> There [are] reports of success with nets and the new medicine, artemisinin, a Chinese Drug made from wormwood against malaria globally.

The story gave the results of an extensive international study on the recent successes in treating malaria. Ancient peoples used a drug to treat malaria; we are just returning to it and call it a wonder drug. The World Health Organization for Global Fund, the primary financial agency for combating malaria, reported that between 2006 and 2007 malaria's mortality was cut significantly. Rwanda, for example, had more than a 50 percent reduction from almost 80,000 deaths in 2006 to somewhat fewer than 40,000 in one year.[72] Children are highly vulnerable to the parasitic infection; a child dies of malaria every thirty seconds.[73] The drive to combat malaria had two fronts, the combinations of which enabled the dramatic global outcome: distribution of insecticide-impregnated mosquito nets and the use of a newly rediscovered herbal medicine, artemisinin, from at least some species of the artemisia genus. Artemisia has received considerable attention in recent years for its effectiveness in combating malaria, most especially in children but including people in all age groups. "If we can scale up like this [with mosquito nets and artemisinin] we should be able to eliminate malaria as a major public health threat in many countries," stated Dr. Michel Kazatchkine, Director of the WHO agency. Forty percent of the world's population is at risk for malarial infection, thus an achievable goal of elimination would, indeed, be great news. The *New York Times* article noted that artemisinin was a Chinese medicine from wormwood, one of the Artemisia species. Virtually unreported is the fact that the ancient Greeks used the plant as an insecticide against mosquitoes; artemisia was employed to reduce fever, as in malaria, and mosquito nets were also used. Why does only ancient Chinese herbal medicine receive the historical credit? As with the pomegranate and mandrake, the answer is not simple but, when the story is unraveled, the plant of Artemis demonstrates a largely hidden but important influence in history, one that only now we can appreciate.

MALARIA IN ANTIQUITY

Malaria is a composite of various diseases caused by different parasitic protozoa carried by certain mosquito species and transmitted through a bite. Four species of the genus *Plasmodium* infect humans with malaria, with only three endemic in the ancient Mediterranean area: *P. malariae* (diagnosed and called quartan fever in the classical medical writings because of periodicity of the fever every third day); *P. falciparum* (called tertian fever with intense fever on the second day of infection and recurring every other day).[74] Complaints are fever, headache, jaundice, anemia, aches and pain, and diarrhea.[75] Because of the periodization and similar symptoms with other diseases in their initial stages, a missed or delayed diagnosis easily ensues. The spleen and liver enlarge, an occurrence observed by classical physicians (as we shall see below). Today, it is called malignant tertian fever. A third agent is *P. vivax* that the ancients probably described also as quartan fever. The term quotidian (from Latin for daily because it spikes every twenty-four hours) is usually applied to *P. vivax* and *P. falciparum malariae. P. vivax* causes a less dangerous and debilitating disease, whereas *P. falciparum* is devastating to the patient and potentially fatal. *Plasmodium falciparum* is the predominant parasite. Today, *P. falciparum* causes more than 120 million clinical cases and more than 1 million deaths occur in the world each year, approximately 80 percent being children less than 5 years of age.[76] One type of tertian fever (*P. vivax*) is, as we saw in chapter three, sometimes called benign malaria (or benign tertian) but only relative to *P. falciparum* infection, because its effects are less serious. Celsus, the Roman author, spoke of two types of tertian fever, one starting and ending just like quartan fever, disappearing for one day and returning on the third, hence tertian. The other type of tertian fever is called "malignant malaria"; Celsus and the author of the Hippocratic work, *Epidemics*, called it "semitertian," the type of malaria today we attribute to *P. falciparum. Epidemics* called the first type "the least fatal and least difficult of all" and the other as "more fatal than any other."[77] Celsus said it was "far more pernicious."[78] To summarize, *P. malariae* causes fevers that recur at approximately three-day intervals (hence, quartan fever), longer than the two-day (tertian) intervals of the other malarial parasites caused by *P. falciparum* and *P. vivax.*[79] The terminology used by the ancients persisted until the nineteenth century. The point is that the ancients could diagnose malaria. One of the earliest records of its devastation comes not in text but on a coin from the Greek city-state of Selinus.

Selinus' Coins

Selinus was a colony in Sicily of the Greek city of Megara in the mid-seventh century BCE. Named after the Greek word for celery (presumably because its wild kind grew well in the region), the city Selinus (or *Selinos*) had a measure of prosperity in the fifth century BCE, as witnessed today by its surviving Doric temple. Something bad befell Selinus' citizens, sapping the energy of many survivors: an epidemic of malaria. The philosopher Empedocles (d. *ca.* 430 BCE), famous for his theory of the four elements (earth, air, fire, and water), supposedly saved the city, according to his ancient biographer: "Since a plague afflicted the people of Selinus because of the evil smell of the adjacent river, causing death and making pregnant women miscarry, Empedocles had the idea of bringing two rivers there at his own expense. By mixing the waters he sweetened them."[80] The major river was named Hypsas (modern Belice River) and the other river, Mazarus. As we saw, the Hippocratic *Airs, Waters, and Places* explained the syndrome of symptoms associated with malaria as caused by "bad air/ *mal aere,*" although that term itself was not used until the fifteenth century. Those cities situated in "marshy, standing and stagnant [water] must in summer be hot, thick, and stinking, because there is no outflow... The malady is endemic both in summer and in winter." Such conditions the Hippocratic writer associated with fevers and, among the symptoms, "large, stiff spleens."[81]

On or about 467 BCE Selinus issued a beautiful didrachm coin to commemorate the salvation of the city through the draining of the marshes, attributed to Empedocles (see figure.4.1). With Selinus' name written on its face, the coin has Artemis driving a chariot; Apollo, the god associated with healing and to whom the Hippocratic Oath was taken, stands behind her, shooting an arrow. According to I. A. McGregor, Apollo discharged "arrows to slay the pestilence, while Artemis represents the women of the city who suffered so severely from the fever."[82] The depiction is interpreted as being Selinus' success in counteracting malaria by draining the marshes. Artemis's presence has been interpreted as representing Selinus' women. There is, I submit, a better reason: her herb was employed to treat malaria. The coin's artistry and its connection with medicine caused the London School of Hygiene and Tropical Medicine (University of London) to adopt the image for its seal. The obverse of this same coin is no less dramatically and well depicted, but its interpretation varies (see figure 4.2). The inscription is clear and to HYPHAS (in Greek letters), the Selinus river god offering *pinakion* (plate) as a votive offering on the altar while in the foreground stands a rooster associated with chthonic (or earth) deities. The representative of Hyphas, an athletic youth, holds a sprig of celery (the city symbol) but it could easily be artemisia. The bull in the background was for sacrifice. Above the bull is a palm sprig. All of these representations are seen as a sacred celebration of the salvation of the city from malaria.[83] Some scholars suggest that the youth also represented Empedocles. Another issue (dated 460–440 BCE) is similarly arranged, but the altar has a snake coiled around it, a medical symbol of Aesculapius (instead of the cock on the altar's top) and, in place of the bull, there is a marsh bird, indicating the connection between the dreaded disease and the pestilential waters from which it came.

Figure 4.1 Selinus coin depicting Artemis driving chariot and beside Apollo shooting arrows to defend city against malaria.

Source: Courtesy of British Museum.

Figure 4.2 Obverse of Selinus coin.
Source: Courtesy of British Museum.

Selinus' coins confirm what we already knew: malarial infections occurred in antiquity, and the deadly *P. falciparum* strain was part of these afflictions. We know that the ancient Greeks associated malaria with marshes and mosquito-breeding areas. Did the Greeks and other ancient peoples know to use and how to use the artemisia plants for malaria? The ancient Chinese had made that discovery.

ARTEMISININ, *ARTEMISIA ANNUA* L., AND MALARIA IN CHINA

The Yellow Emperor's Classic of Medicine or *Neijing* is attributed to the Emperor Huang Di ("TheYellow Emperor"), who ruled about 2700 BCE. Malaria is described in considerable detail—indeed, more detail than our empirical observation would merit because its effect on each organ or body section is described: the account begins with the foot and included is the spleen. For example, in the text this question is raised about a malarial condition: "Can you tell me about the malaria condition that flares up every two days or on some regular basis?" Answer: "This type of skipping days between flare-ups occurs because the pathogen and *wei qui* [roughly 'energy'] engage at *fenfu* [an acupuncture site]; sometimes they meet, however, only every two days."[84] The English translation here employs the term "pathogen" that makes the text seem more modern than merited. Some twelve different "malaria conditions" are delineated. I shall not attempt to make them congruent with modern diagnoses. Acupuncture was the primary therapeutic address for malaria, but there is the intriguing statement: "If the pulse is large, slow, and deficient, one should use herbs but not acupuncture."[85] There we have it and nothing more about the identification of the specific herbs.

Liu An (d. 122 BCE) wrote a pharmaceutical guide in which he named An Lu Zi, a species of artemisia said by the translator to be *Artemisia keiskeana* Miq., as treating blood stasis, abdominal distention, "persisting heat," and pain, but used frequently it makes the "body light," prolongs life, and prevents senility.[86] I only wish that I could confirm these attributes. Translators and scholars of traditional Chinese medicine have not associated "persisting heat" with malaria or even fever, and neither shall I. The earliest confirmed use of artemisia for malaria came approximately 300 years later.

Qinghao, a species of artemisia, was prescribed in a treatise, named *Fifty Two Remedies*, which was found in the Mawangdui Tomb (dating second century BCE). *Qinghao* appeared in a prescription as a fumigant for hemorrhoids, and the text says, "The Jing name of *qinghao* (wormwood) is *qui*."[87] Ge Hong, who lived in the East Yin Dynasty and wrote in 340 CE, said that *qinghao* was good for malaria, if, as probably is reasonably correct, "intermittent fever" is equated with malaria. He wrote: "Another recipe [for intermittent fever]: *qinghao,* one branch, take two *sheng* [2 x 0.21 liters] of water soaking it, wring it out, take the juice, ingest it in its entirety."[88] Chinese scholars claim that *qinghao* was and still is a traditional remedy for malaria in their medical delivery.

As part of Chairman Mao's campaign against Western societies and values, he renewed emphasis on traditional Chinese medicine.[89] As part of that policy, in an anonymously printed paper, the *Chinese Medical Journal* in December 1979 reported a compound found in *qinghao* that subsequently is called artemisinin from *Artemisia annua* L.[90] As best I can tell, the process went like this: artemisinin comes from *Artemisia annua*, not readily found in other artemisia species, and, since traditional Chinese medicine employed artemisia (*qinghao*) for malaria, Ge Hong and other historical medical practitioners must have used this species for malarial treatments. That said a later Chinese medical authority, Shen Gua (1031–1095), said that there were two kinds of *qinghao* and a third kind, identified by modern translators as *Artemisia apiacea* L., produces the most effective means of treating malaria.[91]

MODERN SCIENCE REPORTS

In the 1960s, because of its traditional use, Chinese researchers tested *qinghao* as a malarial treatment, but these series of tests failed to produce positive results. A reason was likely that artemisinin is not soluble in water or ether. Chemically artemisinin (also called in Chinese *qinghaosu*) is a sesquiterpene lactone endoperoxide.[92] The mechanism of action shows that "artemisinin *and its derivatives*" [emphasis supplied] are "toxic to malaria parasites at nanomolar concentrations,"[93] or restated: artemisinin has rapid schizonticidal action—it kills the little parasites and it does so for both *P. falciparum and P. vivax*-type malarias.[94] Whereas the precise pathways are unknown at this time the indications are that the parasites are attracted to the toxin taken up by the bloodstream. Synthesis of the compound from the plant is tricky and its manufacture from the plant is expensive. Because of the dramatic results in treating malaria, research concentrates on economical synthesis procedures, including synthetic manufacture. Promising at present is production of artemisinic acid in engineered yeast.[95] Since artemisinin is not water-soluble— the method given by Ge Hong—, the virtual certainty is that the ancient use of the antimalarial drug from artemisia was a crude extract and, further deduced, the crude plant drug has positive effectiveness to treat malaria. Because *Artemisia annua* L. is a prominently-found species in China, universally the credit for its malarial effectiveness is given to the Chinese. Also, keep in mind that *Fifty-two Remedies* (the earliest recorded use of artemisia in Chinese medical literature) gave two kinds of artemisia, one being *qinghao* and the other *qui*. Also, remember that the modern chemical study showing artemisinin *and other derivatives*' effectiveness in combating malaria. In other words, the various compounds in various species of artemisia may act symbiotically to produce bioactivity in malarial therapy.

Questions Asked

A generalization in plant chemistry is that plants related in the same genus have similar chemistries, although variations occur in different species and even within the same species depending upon morphological sites, methods of extraction, and soil and climatic conditions. Major questions need investigation: (1) Do other species of artemisia possess antimalarial properties? (2) Were the ancient Greeks (such as those in fifth century BCE Selinus) and other Western premodern peoples using *A. annua, A. apiacea*, and *A. lanceolata* species or other species to combat malaria? (3) Did these premodern people use a variety of artemisia species in different ways to prevent mosquito bites and malarial infection?

Delaying a discussion of the various species of artemisia, we find that the immediate question becomes whether the crude drug in a water solution could have positive results in treating malaria. Fourteen healthy male volunteers took one liter of tea made from nine grams of *Artemisia annua* leaves; this is approximately the amount that a traditional herbalist would give for a severe disease. The volunteers' blood samples in laboratory analysis found 94.5 milligrams of artemisinin absorbed. This is 19 percent of the recommended daily dosage of the purified artemisinin.[96] The chemists concluded that the "the possibility existed that bioavailability may be higher from the tea preparation."[97] Ge Hong used the Chinese verb (*jiao*) translated as "wring out"; Elizabeth Hsu proposes that the technical process conveyed by the verb "may have produced an emulsion conductive to the extraction of artemisinin."[98]

Around four hundred species of the broadly distributed Artemisia genus are identified and more than sixty are used in traditional medicine, but the active chemicals are largely unexplored.[99] According to current medicinal chemistry research, artemisinin is found only in *A. annua, A. apiacea*, and *A. lanceolata* species. Specifically negative results for finding artemisinin are reported by Chinese and American chemists who tested *A. vulgaris, A. ludoviciana, A. pontica, A. arbuscula,* and *A. dracunculus*.[100] The educated supposition is that annual artemisia species have a greater presence of artemisinin, but even yields from *A. annua* are relatively low (0.01%–0.08%).[101]

Artemisia Plants (Exclusive of Artemisinin) Have Antimalarial Properties

Recently several laboratory trials show that a number of artemisia species demonstrate active bioactivity against *Plasmodium falciparum*-infected cases of malaria.

Eight plants from Zimbabwe were tested in one study; *Artemisia afra* was found to be the most active as an antiplasmodial agent. When a number of flavones and sesquiterpene lactones, taken from *A. afra*, were isolated and tested and none had strong bioactivity but, when combined as in a crude drug, the plant extract "showed extraordinary antiplasmodial properties." This led the investigators to conclude that "Obviously, the activity of *A. afra* is due to the complex mixture of substances which might act additively or even synergistically."[102] Artemisinin was not found in this plant. Another *in vitro* study of southernwood (*A. abrotanum* L.) showed "appreciable activity" as a potential antimalarial drug. These investigators found no artemisinin presence and, isolating specific compounds had less pronounced results. They concluded that "there are other antimalarial compounds present in these [crude extracts from southernwood] which have not been characterized."[103] The studies cited earlier were all *in vitro* experiments. A team of Indian researchers at the Malaria Research Center in Delhi examined three species of artemisia (*A. japonica, A. maritima, A. niligarica*) by injecting extractions into mice infected with a malarial strain and conducted the study for sixty days. Varying amounts were given to different control groups. The group with the highest concentration had a high fatality rate, presumably because of toxicity. The moderate group had "specific antimalarial activity of different herbal extracts"—all artemisia plants.[104] The control group without any intervention had a greater mortality rate.

Whereas the pharmaceutical community has expended considerable time and money growing *A. annua* and painfully extracting artemisinin, there is evidence that a number of artemisia plants are potentially effective as a drug against malaria. This conclusion can be reached from the scientific literature alone.[105] Add to the scientific studies the fact that artemisia plants have historically been employed over centuries, millennia even, to combat malaria, and we have a reasonable conclusion that our ancestors may have known more than we do about some things about malarial—and without microscopes to see the protozoan. As we shall see the species most frequently used was absinthe (*A. absinthium*), which, according to my research, has not been tested *in vitro* or in animals. Strangely, we do not look where history tells us to look. Before we look at absinthe, we shall explore other ways in which artemisia plants have prevented malaria, namely mosquito nets and insecticide ointments.

Artemisia, Fevers, and Malaria in Western Medicine

Already we have seen some classical medical writers (notably Celsus) and possibly premodern writers who recommended artemisia plants for fever, some giving it as a post-fever symptom for quartan fever, a form of malaria. The Hippocratic author of *Diseases*, discussing a raging fever that remits and reoccurs (a characteristic of malaria), wrote "give a medication, but only on the ninth day; for if you give one when the fever first begins, then after the patient has been cleaned out [with a laxative?] the fever will resume and another medication will be required."[106] The author's precise meaning about periodic fevers that recur on the fifth, seventh, or ninth day is not sufficiently specific for modern equivalences. Seemingly the author thought his readers would know,[107] but the context and description correspond to malarial infections. The significant point is that the ancient physician was not treating the fever itself, but allowed the body to seek its own defenses. Only after a period of disease progression was medication employed. The Hippocratic *Diseases* is a diagnostic work, not a therapeutic handbook, so little in the way of intervention is related. This reopens the question as to whether the ancients were aware of artemisia's benefits in treating malaria pathologically and not just making the patient more comfortable with symptomatic therapies. Already we saw that at least some artemisia species act pharmacologically to attack the protozoa in evidence. Artemisia can be a cure rather than simply countering an unpleasant symptom; malaria is a very pernicious disease.

Over a century ago W. H. S. Jones associated endemic malaria with the decline of ancient civilization and the "fall of Rome." Others (such as Karl Beloch and P. A. Brunt) saw a more direct, but related cause: population decline—"crisis in manpower."[108] In our time approximately 100 million people are annually affected by malarial illness, with about 1 million deaths. The ancients' numbers would have been less (fewer people) but proportionally, malarial conditions would have been significantly worse in antiquity than in today's Mediterranean areas. Approximately in 1513 the Jesuits, accompanying the conquistadors in South America, reported back to Europe an exciting new drug for treating malarial fevers: a bark from a tree called the cinchona. The excitement about the drug from the bark led to a new name for it: "Jesuitenpulver" or Jesuit Powder.[109] Quinine, the drug from cinchona, became the drug of choice for malaria therapy until the parasitic protozoa developed increasing immunity to it. Now we are returning to what earlier was a malaria treatment. Ancient and modern Chinese medicine provides considerable insight into malarial treatments.

A Syriac herbal (*ca.* 600) included artemisia in a prescription that is "good for fever."[110] Similarly Walafrid Strabo said, "Its [wormwood's] powers are famous, its effectiveness proven. It tames a raging fever."[111] Did early medicine employ the plant for the scourge of their day, one whose symptom was a raging fever, namely malaria, which sapped the energy of millions and led to the death of a number of those bitten by the inflicting mosquito? The hypothesis about malarial fevers is that the body raises its temperature to produce a less favorable environment for the protozoa infections. A febrifuge (something that lowers temperature) would, in that case, be counterproductive if all that medicine did was lower the temperature.

Far more common in the early medical accounts was the prescription for various artemisia plants to reduce a swollen spleen, a common malarial symptom. Al-Kindī (d. *ca.* 870) employed absinthe to reduce a swollen spleen.[112] Ibn Sīnā (d. 1037; Avicenna in Latin), the great Muslim physician, recommended absinthe juice for a fever.[113] Macer's Latin herbal said that southernwood "cooled a fever." Absinthe, on the other hand, comforts the stomach but was not given for a fever. Constantine the African's (d. *ca.* 1085) Latin adaptation of an earlier work in Arabic failed to list absinthe for fevers, quartan, or tertian, but he said that it was good for "hard spleens," another way of possibly pointing to the malarial symptom.[114]

Known for the application of alchemy to pharmacy, Paracelsus (d. 1541) is said to have recommended absinthe for malaria therapy, thereby introducing into our study a complex but very important use of Artemis' plant.[115] Earlier in the same century, Pierre Baquillier's commentary on Macer added that both southernwood and absinthe cooled fevers that were rigorous and "horrific fever/ *horripilationes*,"[116] a description that points to malaria.[117] Probably Baquillier thought that Macer meant malarial infections. So also did Nicholas Culpeper in 1653 state specifically of absinthe: It "cures quotidians and quartans."[118] Quartan fever was the mild form of malaria and quotidian fever may have been malaria. Also Culpeper said that absinthe "helps" obstructions of the spleen.[119] Two pertinent points: one, in reference to malaria, Culpeper used the word "cure," a seldom-employed verb in any respectable herbal medicine guide and, two, he prescribed only absinthe species of wormwood and not other kinds (namely Roman or *A. pontica, A. maritima,* and *A. serippium*), which he also discussed.

Summary: Artemisia as a Malaria Drug in the West

We do not know and probably can never know whether classical and medieval peoples had access to the species of artemisia that we are now using to extract artemisinin, today regarded as the drug of choice for malarial infections. We do not know whether crude drug extracts from absinthe, mugwort, and southernwood species-types had bioactivity for treating malaria. We do know, however, that some of these premodern peoples gave the artemisia plants for fevers, and, in some references (above), especially for one type of malaria they called quartan. As we noted in chapter three on the mandrake, ancient medicine did not appear to treat malaria at the first symptoms and fever per se. Indeed, Dioscorides said that absinthe wine is given to those with spleen disease, for the stomach, and edemata, *provided that the patient does not run a fever.*[120] Ancient medicine (at least, so say the medical documents) did not treat a fever but administered drugs toward the end of episodic periods. We know that ancient Mediterranean, medieval, and Chinese peoples employed artemisia plants to combat fever, thought to be malarial, and we have only recently returned to the practice of using artemisia for its pharmacological qualities. Dioscorides's admonition not to give absinthe wine for those with fevers appears simply a realization that absinthe wine is so strong (as shall be discussed below) that it should not be given for diseases with fever. Also the ancients often reported that artemisia plants help a diseased spleen. An enlarged spleen is a symptom of malarial infection. As we saw above, Constantine the African and others prescribed absinthe for a "hard" spleen, thereby raising the possibility that later medical authorities were treating this malarial symptom. It seems certain that the ancients employed artemisia plants; they knew how to use mosquito nets and the plant's juice as a mosquito repellant. Artemisia was a once and future drug. Why, if they had an effective treatment, did the therapy seemingly wane? This question I have often repeated to myself, as when I studied ancient chemotherapy treatments for cancer, only to learn that they employed what we now take as drugs for cancers, but the practice of drug treatments waned beginning in the fourteenth century. Surgeons were successful in persuading their medical colleagues that cancer lesion removal was preferable to chemicals.[121] As we shall see below, the French medical corps gave absinthe both as a prophylactic and as a cure for malaria when their armies and navies served in malarial areas. There was continuity in Western medicine using artemisia plants for malarial treatments, from antiquity to the nineteenth century. We need to look backward until we look forward.

Western medicine employed artemisia species to treat malaria. While we do not know whether the Westerner's crude drug was made in part from the same species as *A. annua* L., the species from which artemisinin is currently taken, we can be reasonably sure that western physicians employed other species (southernwood, mugwort, absinthe) for malaria. Whether this was a widespread practice or infrequently used, we are uncertain—an uncertainty that is true with most ancient drugs until we get to the later Middle Ages and have shop apothecary lists. The Chinese claim, however, that their ancestors first discovered malaria's "cure" is no more solid than the Western case for the prized claim. In the final section on absinthe and alcohol, we shall see how the practice of using artemisia for malaria was common in Western Europe for those traveling or living in malarial areas. There is another claim that Western medicine can make in addressing malarial infections: the mosquito net.

Mosquito Nets

The *New York Times* article quoted earlier said that we are making dramatic strides in malaria treatments by the combination of two approaches: artemisinin and mosquito nets.[122] One is as a drug for those afflicted, already discussed, but the second was something as simple as a mosquito net. In addition to artemisia as a therapeutic agent for malaria, the plant was used as a prophylactic to prevent mosquito bites. The ancients knew to take preventive measures. This is startling inasmuch as the western European and Mediterranean peoples did not suspect the mosquito as the causal agent for malaria, or so say modern scholars.[123] I believe that the preponderance of evidence indicates that likely they did make the association. As already observed, these ancient peoples believed that malaria was a result of pestilential air connected with stagnant water. Federico Borca assembled the numerous references to the ancients' attitudes and fears of marshes.[124] Vivian Nutton, a leading medical historian of classical antiquity, after reviewing the numerous source references, concluded that "These comments on marshes and fens, on stagnant pools and slow-moving rivers, were standard among Greek and Roman doctors of all persuasions, and, indeed, among all writers."[125] They sought to drain swamps and to live, insofar as possible, away from those areas (see the Selinus coins). Major problems occurred with environmental factors, as when the landscape changed near urban areas, thereby increasing "pestilential air" in a slow process. They also sought to protect themselves from insect bites, knowing them to be unhealthy. Dioscorides, for example, said of absinthe: "It keeps mosquitoes away from the body when rubbed [on] with oil."[126] Fascinatingly, he added that absinthe also has other insecticide uses as pest inhibitors: kept in clothes chests, it keeps moths away; put into ink, it keeps mice from eating the papyrus.[127] Numerous modern science laboratory and clinical studies verify that compounds found in absinthe, mugwort, and other artemisia species have active insecticide activities, including fleas, mites, and mosquitoes. [128] Gardeners know to plant wormwood in proximity to other herbs because it keeps insects away.[129] Moreover, the plant's juice is an active mosquito repellant. A compound found in greatest concentration in absinthe is called β-thujone, which is thought to be a major contributor to the plant's insecticide and pesticide activities.[130] (Shortly, we shall see the activity of this compound when ingested by humans in causing a mind-altering experience.)

Herodotus, the historian who gave us the graphic picture of prostitution in Babylon's "love temples," describes the use of mosquito nets in Egypt (which he visited):

> The Egyptians who live in the marshes use for the anointing of their bodies an oil made from the castor berry [from *Ricinus communis* L.], which is known among them by the name of *kiki*.... The contrivances which they use against gnats [mosquitoes] wherewith the country swarms, are the following. In the parts of Egypt above the marshes the inhabitants pass the night upon lofty towers, which are of great service, as the gnats [mosquitoes] are unable to fly to any height on account of the winds. In the marsh country, where there are no towers, each man possesses a net instead. By day it serves him to catch fish, while at night he spreads it over the bed in which he is to rest, and creeping in, goes to sleep underneath. The gnats [mosquitoes], which if he rolls himself up in his dress or in a piece of muslin, are sure to bite through the covering, do not so much as attempt to pass the net.[131]

Greek *kōkōpos* and Latin *culex* are often translated as gnat (also in Latin meaning "midge"), but modern philologists say that the words specifically designate mosquitoes.[132] The use of mosquito nets must have been widespread throughout the ancient Mediterranean world. Horace, the Roman poet, when describing the Roman legionary camp with its standards glowing in the sun, mentions also the distracting view of mosquito nets (*adspicit conopium*) drying in the same scene.[133] Thus far, we have established that the ancients used artemisia plants as insecticides and repellants and that they used mosquito nets to prevent bites. Paulus Silentiarius (sixth century CE) directly tied the mosquito net to Artemis (or, at least, the goddess of love) when he wrote: "I am hung round wealthy bridal beds and am the net, not of the huntress Artemis, but of the tender Queen of Paphos [i.e., Aphrodite]. I cover the sleeper with many a meshed-web, so that he in no way loses the life-giving breeze."[134] The poet, exercising his license, was pointing to Aphrodite as the protector against mosquitoes, not to Artemis. The fact that he employed his substitution of Aphrodite is contrary to the conventional wisdom of his and earlier times was that Artemis was the protector. Paulus also wrote of mosquito nets that preserve people in contrast to the nets to catch fish or birds because these nets permit a safe nap, thereby saving slaves from having to fan away flies (*musca*).[135] Propertius metaphorically spoke of mosquito nets when speaking of Cleopatra's ambitions on Rome to drape the Tarpeian rock, Rome's symbolic equivalent of the American Statue of Liberty.[136] Even though there is no explicit statement that ancient peoples put artemisia juice on the nets themselves, there is ample evidence that they used the juice as a repellant on their bodies. Besides, there is no need to put the juice on the net because, as explained by Agathias Scholasticus (a Greek poet of the late sixth century): "Not a single mosquito, however small, will manage to get through the fabric of my net."[137] An agricultural compilation, known as *Geoponica* (tenth century), lists some of the plants used to keep away or kill insects, notably mosquitoes, as fumigants or sprinkled around the house, rubbed on the body, or placed in bedding.[138] Even with the scattered evidence presented by the surviving classical sources, this is a reasonable conclusion: ancient peoples used insecticides in a rational way (as defined by modern science), and they were aware of the dangerous mosquito.

How widespread was this knowledge? The preponderance of evidence indicates that knowledge about mosquito nets and use of insect repellants was common. As another indication of the connection between gods and herbs, a medieval practice associated with St. Luke was to make an application out of wormwood (absinthe) and other herbs burned to a powder over a slow fire: mixed with honey and vinegar, anoint on the body at bedtime, and say three times: "St. Luke, St. Luke, be kind to me, in dreams let me by true-love see." The anointment was a poetic expression for preserving health from mosquito bites so that the person could live to dream of his beloved.[139] In 1577 Thomas Tusser penned this verse:

> While Wormwood hath seed get a handful or twaine
> To save against March, to make flea to refraine;
> Where chamber is sweeped and Wormwood is strowne,
> What saver is better (if physick be true)
> For places infected than Wormwood and Rue?
> It is a comfort for hart and the braine,
> And therefore to have it is not in vaine.[140]

Tusser's singling out the wormwood to prevent flea infestation could possibly be a precaution against the bubonic plague although this is a guess. Writing in England, Tusser did not fear the malarial mosquito, but he praised wormwood insecticide's comforts. As a repellant, wormwood and other artemisia plants must have been inexpensive, considering their ubiquitous range and hardiness. The evidence indicates that not only did artemisia act as a repellant and insecticide, but also the plant saved people from both uncomfortable and deadly bites, as Tusser observed; in malarial areas, the herb preserved their health and life. The deadly qualities of diseases by insects from swamps were known and people used rational measures to save themselves from death or, in the least, from scratching.

Without having explicitly made a link between malaria and the mosquito—at least, in the surviving documents—, ancient peoples took the same precautions as if they knew. While not explicitly stated, the connection was definitely implicit. Varro, a Roman writer, wrote: "Precautions must also be taken in the neighbourhood of swamps both for the reasons given [i.e., 'unwholesome (air) in the summer'], and because certain minute creatures, which cannot be seen by the eyes, which float in the air and enter the body through the mouth and nose and there cause serious diseases."[141] And Columella came closer to a connection between malaria and the mosquito—those tiny things that Varro named—when he said cautioned against building near marshes: "Worst of all is swamp-water, which creeps along with sluggish flow; and water which always remains stagnant in a swamp is *laden with death* [emphasis supplied]. But this same water, harmful though its nature is, is purified by the rains of the winter season and loses its virulence."[142] And, elsewhere, Columella directly ties these insects to deadly diseases when writing about the desirability to locate buildings away from marshland:

> [A building near a marsh] throws off a baneful stench in hot weather and breeds insects armed with annoying stings, which attack us in dense swamps; then too it sends forth plagues of swimming and crawling things deprived of their winter moisture and infected with poison by the mud and decaying filth, from which are often contracted mysterious diseases whose causes are even beyond the understanding of physicians.[143]

Clearly Columella connected the "decaying filth" of the swamps to the bites of insects, and these bites were potentially fatal. The swamps were "laden with death," he said. The nature of ancient documentation does not permit us to know every aspect of daily living, but there is sufficient evidence that some ancients connected mosquito bites with a deadly disease. Certainly they (at least some of them) knew that it was not the pestilential, corrupted air per se, but an insect or tiny, invisible organism that caused the dreaded disease. The ancients were ravaged by malaria. Their ancient writers clearly stated that mosquito bites or stings were potentially fatal. Reason causes us to conclude that the ancients (at least some of them) knew that small things stinging or biting them caused malaria. For those reasons they attempted to avoid marshy swamp lands, to drain stagnant waters, to seek elevation where mosquito would not fly in higher altitudes, and to take measures to protect themselves by using artemisia oils (as repellants) and mosquito nets. All of these measures would come at some expense in time, money, and deliberate actions; all point to the conclusion that some of the ancients knew the causes of malaria and took rational steps to protect themselves.

The mosquito net and the plant's insecticide qualities enabled many people from having their vital energy and spirits drained by an infectious agent. The recent *New York Times* article (quoted earlier) trumpets the modern measures we are taking through the use of artemisinin and mosquito nets and the hundreds of thousands of lives these actions are saving. Even if the previous scholars were correct in thinking that the classical ancients did not connect the mosquito with malaria—and I think that they did—, the measures they took would have had the same beneficial results.

Malaria and Decline of Classical Cultures

The final question, one that is bothersome but explainable, is why did malaria increase at times in the ancient and premodern worlds and wane at others? Environmental changes are part of the answer, as in Campagna when there were increases in swamplands attributable to a moister climate. The infecting species of mosquitoes do not tolerate a cold climate. The famous Pontine Marshes south of Rome inspired mammoth projects; beginning with Julius Caesar, then Anthony and Augustus, governments undertook to drain the swamps, but total success was not achieved until Mussolini.[144] Modern visitors to Italy will want to visit the ancient ghost city of Paestum, a Greek city-state a short distance off the Autostrada between Naples and Salerno. The ruins of Paestum encompass an entire city with buildings and standing roofed temples. Paestum was abandoned, not because of the Goths or

Vandals, but earlier. The exodus of Paestum's inhabitants began probably shortly after Vesuvius' eruption that blocked streams and caused marshes to infiltrate the lands around the city. Speaking of the volcano, I recall a student's identification for Pompeii when he wrote: "A city destroyed in 79 [CE] by an eruption of saliva from the Vatican." However, what destroyed Paestum was neither saliva nor lava, but the troublesome mosquito. Those pesky bugs were all too determined to find a bite.

Absinthe, Artemisia's Uplifter and Downplunger

An old Anglo-Saxon saying was "to drink *wermōd* [wormwood] in a dream betokens strife."[145] Alcohol and medicine are closer than cousins: many people historically associated medicine with its preferred delivery system in beer, wine, various fermented plants, and, finally, distilled beverages. Ironically the Arabs, forbidden by the Qur'an to drink inebriating drinks, invented distillation and entered into our vocabulary a new Arabic term, alcohol. My father was known to take a shot or two of whisky; he usually referred to the beverage as "his medicine." When I curiously asked the purpose for the medicine, he replied, "It is a medicine and a tonic." And so it was, although I suspected his motives were nontherapeutic, and I did not appreciate his defense. As I learned more about ancient medicine, I realized more the truth of his claims; he was in the historical mainstream in associating alcoholic drinks with medicinal deliveries. For example, when Europeans learned the South American use of cinchona to treat malaria, they realized quickly that its taste was very bitter. The medicine was more palatable when mixed with alcoholic drinks or inert flavoring agents. In the eighteenth century in India, the British mixed cinchona with a newly invented alcoholic drink called gin. Gin was an avant-garde discovery made probably in the Netherlands (seventeenth century) by Franciscus Sylvius, a physician, who distilled juniper berries as a more pleasant and inexpensive way to deliver juniper's diuretic, renal and other medicinal qualities. The English word gin derives from the French "genièvre," meaning juniper.[146] In 1817, a French chemist isolated quinine from the natural bark of the cinchona tree but, alas, its taste was still bitter. Thus, the British in India took their daily prophylactic gin and tonic (quinine). Armed with these combined and pleasant medicines, Europeans were enabled to penetrate malarial areas of Asia and Africa for "discoveries" and colonies. Gin and tonic medicinal drinks were in a long line of alcohol-based therapies. Very pleasant medicinal "spirits" were developed in various monastic pharmacies; Benedictine and Chartreuse quickly coming to mind. Some alcoholic drinks, however, were taken for their medicinal effect to alter the mind even more than alcohol alone.

As the author of the most comprehensive and superior pharmaceutical guide in antiquity, Dioscorides seemed reluctant to include various wines with medicinal additives; he recognized that the action of medicinal wines was the composites of wine and additives. He seemed hesitant about including such wines when he said, "we hesitate to give wine but are forced to do so by the hankering of the patient."[147] The ancients seemed to associate wine (or other fermented drinks) with medicine and Dioscorides was compelled to add the first section of his Book Five on some of the medicinal wines. We can surmise that vendors sold various medicinal wines. From times more remote than our records, artemisia plant drugs were frequently delivered with beer or wine. Already beginning with the Ebers Papyrus we saw artemisia delivered in beer, whereas Assyrian medicine employed fermented plant juice. Dioscorides said, "We disapprove of using it [absinthe] in drinks because it is bad for the stomach and gives headaches."[148]

Alcohol and Absinthe

Despite Dioscorides's disapproval, absinthe and other artemisia species were very much a part of the lives of countless people throughout the ages. Absinthe's benefits were regarded as both medicinal and recreational. As we shall see for the nineteenth century, its recreational use was expanded to include creative and inspirational use. In the ninth century, Walafrid Strabo, a very pious monk and writer about the herbs in his monastic garden, wrote this verse about absinthe (*absinthium*):

> The next bed grows bushes of bitter wormwood. Its supple stem
> Resembles the Mother of Herbs [referring to mugwort, *Artemisia vulgaris* L.] but the leaves have a different color,
> The smell of its downy branches is different, too, and the brew
> It makes has a bitterer taste by far.
> Its powers are famous,
> Its effectiveness proven. It tames a raging thirst; fever
> It banishes. If, besides, your head should suddenly start to
> Throb and throb with pain, if fits of fainting worry you,
> Seek its help:
> Boil the bitter stem of a plant
> In leaf, tip the brew into an ample basin
> And pour it over the top of your head. Then having bathed
> Your soft hair with the liquid make a garland of leaves
> (Do not you forget this) and put it on, so that the bandage
> Gently binds your hair and holds the warmth in it.
> A few hours later—not many—you will be marveling
> At this yet further proof of the healing powers of wormwood.

Walafrid's use as a prepared garland for its "marvelous" effects seems a stretch unless he was referring to a ritualized use discussed above or, possibly it was to keep his abbot away. However, a legacy may be found in modern Mexico, where women celebrate a festival in honor of the goddess of salt by wearing garlands of wormwood as they perform a traditional dance.[149] As an additive to beer, absinthe was employed religious ritual and worship, connected to Artemis, Salt, and other gods from antiquity to the present.

Walafrid began his discussion of absinthe's medicinal qualities by saying that the herb imparted a bitter taste to beer. For a long period, wormwood in its various species was added to beer to make its taste more pleasant. Not until the ninth century was when the use of hops invented in Germany—where else! Gradually hops replaced absinthe as a beer additive. According to modern tastes (and health), hops are superior to wormwood, although I speak as no authority. In 1710, in England, the use of absinthe in beer was forbidden for brewers' use; however, retail and local brewers were allowed to continue their own additives, including absinthe, since people preferred familiar tastes in their ales. A winter brew called purl was permitted when orange peel and wormwood (absinthe?) were mixed in the recipe and became a working-class favorite. The *London and Country Brewer* (1759) was no purl supporter when it wrote: the "common weedy Wormwood . . . which Compound one would think, more fit for Puke, than a grateful cordial, stomachic Bitter."[150] In the reign of George III (1760–1801) purl beer with absinthe was a "very popular early morning drink" and called a "hot beer with a dash of gin."[151] What a way to begin a day! No wonder the American colonies were lost.

From the Egyptians down to the present day, many of us have tasted absinthe in alcohol in a mild preparation. In 1786, in Turin, Italy, Antonio Benedetto Carpano experimented with the German recipe for wormwood-fortified wines. Camparo is a famous brand of vermouth widely sold, especially in Europe, taking the name from Gaspare Campari who marketed the formula between 1862 and 1867. The name vermouth is an Italian adoption from the German *wermut*. In the United States, the popular use of vermouth is to mix with gin or vodka for martinis. French vermouths today are usually white and dry whereas Italian ones are darker and sweeter, a subject for which I am slightly more of an authority.

Absinthe's Impact on Nineteenth Century

By the sixteenth century, distilled liquors were a part of most Europeans' lives. Some liquors served both as medicines and for their narcotic and/or mind altering qualities. Often they were regarded as an elixir, the quasi-mystical substance that would eradicate disease and aging, a subject in our last chapter. Prominent among the new elixirs was what Paracelsus and others referred to as *absinthe Suisse*.[152] Switzerland was deemed the best producer of absinthe-laden alcoholic beverages that served as medicine. In 1792 Pierre Oridinaire, a French physician, fled from the Revolution in Paris to settle in Couvert, a Swiss village. He developed his own formula for absinthe alcohol, which he enjoyed entirely too much as he died of its cumulative effects. Nearby were the Henriod sisters, one of whom was Oridinaire's housekeeper, who made the absinthe drink—whether by the same formula or a variant is debated. Controversy arises about whether the sisters received the formula from Oridinaire or whether they made up the story to lend prestige by the association with a medical doctor. This much followed: the sisters gave the recipe to a Major Daniel-Henri Dubied, who appreciated the beverage and drank it twice daily—it stimulates the genitals, he said! In drinking absinthe in the mornings and evenings, Dubied was following the advice of Nicholas Culpeper, who in 1653, said: "This infusion [of absinthe], drunk morning and evening for some time, helps hysterics, obstructions of the spleen, and weakness of the stomach."[153] Judging by the evidence, daily absinthe-drinking must have been a common practice for some centuries. What had changed was distillation thereby making concentrated absinthe oil with distilled alcohol more potent. Dubied passed the formula for his daily tots to Henri-Louis Pernod, his son-in-law. Pernod marketed the drink, and thus began the famous Pernod Fils Company in 1805, a company still selling beverages (see figure 4.3). *Pernod aux extraits de plantes d'absinthe* also was flavored with fennel,

Figure 4.3 Posture (19th c.) Absinthe Vichet by Nover.

Source: Courtesy of David Nathan-Maister and the Virtual Absinthe Museum, http://www.oxygenee.com.

melissa (a mint today used for mild depression), and anise; its imbibers thought that it attacked or prevented fevers, helped digestion, and enhanced one's appetite.[154] Away from the malarial mosquitoes, northern Europeans were prepared, just in case of a southern invasion.

ABSINTHE CRAZE

With Pernod and other brands of absinthe on the market, there appears some increase in fashion for the alcoholic drink and medicine as an elixir, compared with its use in previous centuries. The monumental burst of popularity is traced to France's war in Algeria where intestinal parasites and malaria probably killed more soldiers than were killed by the local militias. The army's medical corps issued an absinthe alcoholic drink in place of quinine for malaria. Moreover, it protected against intestinal parasites. Quinine was said to be too expensive, and the absinthe medicine proved popular with the soldiers as some of them transferred from Africa to Madagascar and Vietnam. The *London Times* reported in 1872 that "Fever made grievous havoc in the ranks of the army."[155] With each sip of the medicine wonders were worked for the soldiers' morale: protection from malaria and worms. Don't drink the water; it is unhealthy, and so it was. However, it was dangerous also to drink its medicinal substitute, absinthe alcohol, just as the classical and medieval sources had warned, and moderns neglected to heed. Initially provided by the French army's medical corps, absinthe alcoholic drinks were popular throughout Europe and, to a lesser degree, America, but soldiers were partly responsible for its popular acceptance seen in the 1871 cartoon spoof of a soldier drinking absinthe (see figure 4.4).

The sale of absinthe alcoholic drinks became fashionable in Europe, most especially in France. Like alcohol, it was intoxicating. As a current Web site that describes the several absinthe-drinking rituals, its effects are "visual disturbances, unusual sensitivity to light and color, mild euphoria, and a peculiarly clear-headed type of drunkenness."[156] This modern description is parallel to the earlier Anglo-Saxon quotation: "to drink *wermōd* [absinthe] in a dream betokens strife." Walafrid Strabo's "marvelous" action for a garland made of absinthe is probably an allusion to the medieval ritual discussed above. Rituals and absinthe-drinking are historically linked. When ingested, not worn, absinthe's actions are in the head, not on it. Intoxicating, mind-altering, convulsive narcotic: these words are used to describe its effects, but the drink was devised as medicine.

A manufacturer's manual for the distillation and manufacture of alcoholic liquors published in Paris in 1866–1867 gave a number of recipes for absinthe alcoholic drinks but said these drinks are "a horrible curse, which is killing the youth of our colleges, decimating the army, and will cause fatal debasement of the rising generation."[157] Absinthe was linked as a cause, if not *the* cause, of France's lost Franco-Prussian War, near-loss in World War I, and the "decadence threatening the British Empire."[158] Newspapers and journals of the nineteenth and early twentieth centuries record numerous human tragedies in multiple numbers: families ruined and horrible crimes committed by people crazed on its intoxication.

Absinthe has compounds, especially notably α- and β-thujone (also discussed earlier) that occupy the same receptor sites as marijuana; both seem to have similar molecular structures for this reception.[159] A recent laboratory study using rats,

Septième année. — N° 295. Un numéro 10 centimes. DIMANCHE 21 JUIN 1874.

FONDATEUR

F. POLO

ABONNEMENTS

PARIS

52 numéros.............. 6 fr.
26 numéros.............. 3 —

Les abonnements partent du 1er de chaque mois

BUREAUX

16, rue du Croissant, 16

L'ECLIPSE

FONDATEUR

F. POLO

ABONNEMENTS

DÉPARTEMENTS

52 numéros.............. 8 fr.
26 numéros.............. 5 —

ANNONCES

Fermage exclusif de la publicité

ADOLPHE EWIG

10, rue Taitbout, 10

ADRESSER LES ABONNEMENTS ET RÉCLAMATIONS A L'ADMINISTRATEUR DU JOURNAL

L'HEURE DE L'ABSINTHE, PAR GILL.

Voir au dos les vers de Gill.

Figure 4.4 Giclée art print of Absinthe showing a caricature by Gill, satirizing the French military's love of Absinthe, 1874.

Source: Courtesy of David Nathan-Maister and the Virtual Absinthe Museum, http://www.oxygenee.com.

however, indicated that, while the intoxicating effects of marijuana and thujone were similar, the activation of the same receptors is not supported by the study's data.[160] Both marijuana and absinthe are terpenoids, formed by a similar biosynthetic mechanism, and possess similar molecular structures. An oxygen molecule is the difference and probably occupying the same receptor sites without changing the orientation of the oxygen molecule.[161] Rats and humans are different and may have different results in testing. This much is certain: both substances produce a mind-altering experience or, in more precise terms, it has a psychotominmetic effect. Of the two, thujone is likely more pronounced. Its effects, of course, depend on concentration and amount of alcohol in its delivery and can have a profound mental alteration that leads to physiological impairments: convulsions, vertigo, insomnia, nervousness, nausea, and tremors.[162] The twenty-fifth edition of the United States Dispensatory (1955) said of absinthe in a volatile oil, which is how it was often served especially in the nineteenth century: "an active narcotic poison [that] causes trembling, stupor, and later violent epileptiform convulsions, involuntary evacuations, unconsciousness, and stertorous breathing, which may or may not end in death."[163]

Creativity, Death, and Absinthe Addiction in Nineteenth-Century France

Ernest Hemingway's *For Whom the Bells Toll* (first published in 1940) has this conversation in Spanish civil war between Robert Jordan, his central figure, and a gypsy about a bottle of absinthe:

> "What drink is that?" the gypsy asked.
> "A medicine," Robert Jordan said. "Do you want to taste it?"
> "What is it for?"
> "For everything," Robert Jordan said. "It cures everything. If you have anything wrong this will cure it."[164]

Hemingway's advocacy for absinthe as a medicine, albeit in a fictional character's words, came near the mid-twentieth century. Despite his claim that "it cures everything," Hemingway's absinthe was more for the soul than the body.

Many artists and literati found inspiration and joy in absinthe. Ernest Hemingway continued this exchange between Robert Jordan and the gypsy to whom he offered a drink of absinthe:

> He [Robert Jordan looked forward to] being able to read and relax in the evening; of all things he had enjoyed and forgotten and that came back to him when he tasted the opaque, bitter, tongue-numbing, brain-warming, stomach-warming, idea-changing liquid alchemy. The gypsy made a face and handed the cup back. "It smells of anise but it is bitter as gall," he said. "It is better to be sick than have that medicine." "That's wormwood," Robert Jordan told him. "In this, the real absinthe, there is wormwood. It's supposed to rot your brain out but I don't believe it. It only changes the ideas."

Hemingway came near the end of the period when artists and literati were enamored of the alleged positive affects that absinthe had on the human mind.

In 1859, Edouard Manet finished a large-size, gripping painting of a disgruntled bum with his title for it: *The Absinthe Drinker*, thereby beginning a long list of artists whose subject was either absinthe drinkers, absinthe bottles, or both: Edgar Degas, Gustave Moreau, Henri de Toulouse-Lautrec, Vincent van Gogh, Paul Gauguin, and Pablo Picasso. In 1860, one year after Manet's *The Absinthe Drinker*, Henri Balestra wrote *Absinthe et Absintheus*, known for the beginning of the socialization of absinthe drinking. In Europe, particularly in the urban areas of France and Spain, the etiquette of cafés developed absinthe-drinking rituals. In New Orleans an absinthe cafe still stands. Writers and artists had, or at least, thought that they had, their inspiration derived from the plant's effects; these include Charles Baudelaire, Gustave Courbet, Paul Verlaine ("I get drunk to keep up my reputation"), Arthur Rimbraud, Oscar Wilde, Ernest Dowson, and Emile Zola. While touring America, Oscar Wilde said, "Absinthe has a wonderful color, green. A glass of absinthe is as poetical as anything in the world. What difference is there between a glass of absinthe and a sunset?" Well, there was a sunset in van Gogh's absinthe-drinking life. He had two sustained episodes of depression followed by bursts of high energy, productivity, and creativity. More than 150 physicians have ventured diagnoses, but a consensus of opinion asserts that absinthe played a key role.[165] His bursts of colors are the subject of a recent study that links absinthe with creativity.[166] Van Gogh was so enamored with the drink that he painted it (see figure 4.5). At the age of thirty-seven, driven by a distorted mind, he died of a self-inflicted wound. His mind produced the splendid images on canvas with yellow hues and halo effects, conjecturally derived from a drug-distorted brain.[167] What conversations he and Gauguin must have had when they drank absinthe together in Paris and Arles!

In fact absinthe alcoholic drinks are illegal in most western countries, with legislation that began early in the twentieth century, as we shall discuss later. For now, no reader should drink absinthe drinks (although legal in some countries); that said, absinthe, a species named for the goddess and still known in the Christian world as the Mother Herb, regularly affected many people. Thujone, a psychologically active compound, is a volatile oil and its presence in absinthe accounts range from 0.25 to 1.32 percent of content. Preparation would be by making a decoction, a "cooking down" for higher concentration. It takes only a few drops, as recounted by Robert Service in "The Shooting of Dan McGrew":

> A bunch of the boys were whooping it up in the Malamute saloon; The kid that handles the music-box was hitting a jag-time tune; Back of the bar, in a solo game, sat Dangerous Dan McGrew, And watching his luck was his light-o'-love, the lady that's known as Lou. When out of the night, which was fifty below, and into the din and glare, There stumbled a miner fresh from the creeks, dog-dirty, and loaded for bear. He looked like a man with a foot in the grave and scarcely the strength of a louse, Yet he tilted a poke of dust on the bar, and he called for drinks for the house. There was none could place the stranger's face, though we searched ourselves for a clue; But we drank his health, and the last to drink was Dangerous Dan McGrew. There's men that somehow just grip your eyes, and hold them hard like a spell; And such was he, and he looked to me like a man who had lived in hell; With a face most hair, and the dreary stare of a dog whose day is done, As he watered the green stuff in his glass, and the drops fell one by one.[168]

Figure 4.5 Vincent Van Gogh's "Absinthe bottle".
Source: Courtesy of Van Gogh Museum, Amsterdam.

The multiple drops of "green stuff" were absinthe. A couple drops would have been plenty for most men, but not for Dan McGrew.

SUMMARY OF ABSINTHE AS RECREATIONAL DRUG

From the Yukon back to biblical times, absinthe has been a scourge to humans, conjecturally, nearly equal to its beneficial qualities for mankind. The ratio of benefits to harm strongly favors absinthe and other artemisia species as a blessing to women from the goddess (or God, as in biblical times). The Mother Herb was so named because of its importance for women's general health in a variety of ways. Lamentations (3: 15) is emphatic about the misery wormwood (or absinthe) can cause: "He [God] hath filled me with bitterness, he hath made me drunken with wormwood."[169] God's judgment for apostasy is similar to the results of drinking absinthe, namely a madness and disorientation. Shortly thereafter, see Lamentations (3:17): "Remembering mine affliction and my misery, the wormwood and the gall." An allusion to absinthe's mind-altering qualities was made in Amos (5:7): "O you who turn justice to wormwood, and cast down righteousness to the earth." The Mother Herb was also called "the herb of forgetfulness," because of its mind-altering qualities. An old Russian story tells that mankind forgot the understanding of speech and the uses of plants when a peasant spoke the name for the plant despite an injunction not to do so.[170]

CONCLUSION

In 1905 in Switzerland, a thirty-two-year-old laborer drank absinthe until he was literally out-of-his-mind; he took his rifle and killed his wife, two daughters, and attempted unsuccessfully to kill himself.[171] In the 1890s, nation-states in Europe and North America had begun to regulate substances; at first, it was pharmaceuticals, because of particularly egregious practices by unscrupulous peddlers and manufacturers who adulterated and substituted base substances sold under the labels of legitimate drugs. Across much of Europe and North America came the realization that absinthe was a pernicious and devilish drug that ought to be banned. Succinctly, Raoul Ponchone said, "C'est le diable fait liquide" or loosely, the devil is in the bottle.[172] In 1915, France (where there was perhaps the greatest addiction among the working and creative classes) and the United States banned absinthe alcohol. Many European nations followed; other nations regulated the amount of absinthe. In the 1990s, the European Union reauthorized the sale of absinthe alcohol (vermouth was always excepted). In 2007, distillation resumed in the United States with restricted amounts of absinthe. By 2008 there were more than 200 brands available.

Artemisia plants have a long history. The plant of the goddess Artemis saved women from hardship, pain, and death; it gave comfort and health to many people throughout the millennia. Through the generations, its health benefits were overshadowed, not because the plants' chemistries changed, but because humans abused its compounds. The obverse is also true: because of its antimalarial qualities, it is saving hundreds of thousands lives; as we learn more about its complex chemistry, these numbers should expand into the millions annually. Poor Artemis! She would have been appalled at utility of the Mother Herb while rejoicing about the lives she saved and made healthier. Perhaps that is what Victor Hugo had in mind when he wrote:

> Who brought you into the world, and ever since, a tender mother,
> Standing you double stead in this bitter life,
> Has always drunk the absinthe and left the honey for you.[173]

CHAPTER 5

THE CHASTE TREE

> This is my covenant, which you shall keep, between me [God] and you [Abraham] and your descendants after you: Every male among you shall be circumcised.
>
> —Gen. 17: 10 (Rev. Standard trans.)

> He [Antiochus Epiphanes, King of Seleucid Empire that included Judea, ca. 175BCE] authorized them to observe the ordinances of the Gentiles. So they built a gymnasium in Jerusalem, according to Gentile custom, and removed the marks of circumcision and abandoned the old covenant.
>
> —1 Maccabees 1: 12–15. (King James trans. of Apocrypha)

This all appears strange to us today—not the Jewish circumcision practice but its reversal, or "uncircumcision." Assuming that it was possible, how might they have done it? Answer: the assumption can be erased: they did uncircumcise. The surgical details are given in the ancient medical accounts, but, still, mysteries remain: how could skin grating be accomplished without something to prevent a youth from having, shall we say, erectile function? A requirement would be the opposite action to our Viagra®! This story is untold; one plant had an enormous impact in history. Alas, in modern times the knowledge of this plant is virtually forgotten. This chapter reveals part of that story and the next chapter will tell the plant's reverse role in witchcraft during the medieval period, a story that accounts for our forgetting its history.

CIRCUMCISION AND ITS REVERSAL, *EPISPASM*

In his epic novel, *The Source*, about Israel's history, James Michener has a chapter with the title: "In the Gymnasium"; it is about semi-fictitious events allegedly taking place in 167 BCE in the Greek "gymnasium," which means place for naked play. Therein lay the trouble. The Israelites considered circumcision an irreversible act of religious and cultural identification.[1] Herodotus said the practice originated in Egypt, as well it may have, as a young man is seen undergoing circumcision depicted on a relief found on the tomb of a vizier to the Pharaoh Teti (*ca.* 2300 BCE) at Sakkara, 6th Dynasty (see figure 5.1). In the panel on the right, a surgeon (priest?)

Figure 5. 1 Egyptian circumcision procedure from tomb of a vizier to the Pharaoh Teti (ca. 2300 BCE) at Sakkara, 6th Dynasty.

Source: Photo credit: Werner Forman/ Art Resource, NY.

cuts a young man with a flint knife. In front of the surgeon-priest, the hieroglyphic signs say: "The *hem-ka* priest is circumcising." Reading the writings at the top from left to right, the surgeon-priest says to the attendant: "Hold him fast; do not let him fall," as he cuts with the flint knife. After the surgery, in the second panel the youth says to the surgeon: "Rub it well that it may be effective," to which the surgeon replies: "I am acting for your praise."[2] The "rub it well" phrase points to a regional anesthesia and/or antiseptic.[3] Early in the same millennium as the Egyptian relief, however, circumcised warriors in Syria are clearly shown.[4] The Jews were not alone in the religious surgical observance, but resistance to circumcision came when Judea came under the power of the Greeks following Alexander the Great's conquest. It was truly a cultural clash.

Alexander's policies were followed by his successors, but exactly what they were is open to interpretation. Historians argue about whether Alexander's motives were to promote a fusion of Europe and Asia—most historians now doubt fusion as an intentional policy goal—but the inescapable fact is that Alexander and his successor spread Hellenic culture throughout the vast areas of Asia they visited.[5] Many Asians supported Hellenization, others vehemently opposed it. This cultural clash took two major forms: the presentation of Greek drama, and the participation in sports and with it, the gymnasium. James Michener understood the cultural clash when an athletic Jewish youth, one of his protagonists, was ashamed of his nakedness, not that he was naked but that his glans was showing. Strangely (to us), there was only one particle of modesty among the ancient Greeks: while public nudity was acceptable, the sight of the glans was strictly reprehensible. Some fifth-century Greek vases depicted villains with bulbous externalized glans, whereas heroes or idealized men had tapered, covered glans. Ugly people, barbarians, and slaves were often depicted with exposed glans.[6]

The Greeks abhorred any mutilation of the body; to them, the abhorrence was just as religious as was circumcision among the Jews.[7] Greek boys born with an abnormally short foreskin could go to a surgeon to add length to cover the glans. One might say that this operation, known as epispasm, was a mutilation itself, but the Greeks considered it merely a correction of nature's error. Michener's athlete was encouraged to participate in the regional games at Antioch, but he could not do so being circumcised. Solution: a surgical procedure to reverse the action and replace the foreskin. The drama unfolded when he took his place on the field with other nude athletes and from the crowd came: "Menelaus! You are one of us!" And a chorus of fans added: "He is a Greek! He is a Greek!"[8] Oh, my, how ashamed was his father as he witnessed his son's new nudity in a public arena!

Operation of Epispasm

The name, epispasm (*epispastikos*), is from St. Paul (1 Corinthians 7. 18):

Was any one at the time of his call uncircumcised? Let him not seek to remove the marks of circumcision. Was any one at the time of his call uncircumcised (*epispasthō*= "pull-over").[9]

An unlikely candidate for new medical terms, St. Paul gave us the word epispasm. The Greek (or Seleucid) King Antiochus IV (surnamed: Epiphanes, r. 175–164 BCE) was not anti-Jewish in his policies: he granted the Jews various tax exemptions and significantly subsidized the Temple in Jerusalem. Generally Hellenistic rulers were religiously tolerant, but they drew the line on circumcision, so repugnant was it to them. Antiochus forbade circumcision on pain of a dreadful death.[10] Maccabees 2 (4:12) expressed Antiochus' actions this way: "For with alacrity he founded a gymnasium right under the citadel, and he induced the noblest of the young men to wear the Greek hat." By the second century BCE more Jews lived in Alexandria, Egypt, than in Jerusalem, but in Alexandria all non-Greeks were taxed. Adolescent males (known in Greek as ephebes) exercised at the gymnasia and some Jewish youths, who were not circumcised or had its reversal, sought to cross the line for full citizenship through socialization and participation in athletics in the gymnasia. This caused the author of 3 Maccabees to condemn those Jews who became Greek ephebates.[11] Some Jews, inclined to accept Hellenism, accepted the new law against circumcision of eight-day-old males and reversals for men; a Jewish conservative group refused; some of them were martyred by paying the ultimate penalty. The pro-Hellenic group was later called Sadducees, whereas the conservatives were to be called Pharisees. The issues dividing these two "parties" were present in Jesus's time. The Jewish historian Josephus wrote: "They concealed their circumcision to be Greeks even when unclothed."[12]

There existed an intermediate form of, shall we say, "circumcision treatment" that Soranus described as follows:

> If the infant is male and it looks as though it has no foreskin, she [midwife] should gently draw the tip of the foreskin forward, or even hold it together with a strand of wool to fasten it. For if gradually stretched and continuously drawn forward, it easily stretches and assumes its normal length and covers the glans and becomes accustomed to keep the natural good shape.[13]

This method is called infibulation or in Greek *lipodermos*; this procedure was not at the heart of the issue. Infibulation was simply stretching the skin, which would not fool a close observer and, in any case, would not be successful for all patients. The honey (used as an emollient for thirty days) and the herb giant carrot (*Thapsia garganica* L.) were used in the infibulation procedure well described by a number of ancient authorities.[14] Dioscorides stated: "It [giant carrot] is used also for the foreskin, raising a swelling on people who have no foreskin but not because of circumcision. For the defective part of the phallus is bathed and softened by the ointment, it fills up."[15] Celsus, the Roman medical writer, doubted the effectiveness of infibulation.

A true circumcision reversal procedure was central to the cultural clash issue. A document known as the Testament of Moses (written perhaps during the period of Antiochus IV[16]) regarded Antiochus's edict against circumcision as oppression and torture when the treatise stated: "Their [the Jews] young sons shall be cut by physicians to bring forward their foreskin."[17] This surgical procedure is the heart of the cultural issue, not merely skin stretching. Celsus relates the details of two epispasm surgical procedures; in his words, "If the glans is bare and the man wishes for the look of the thing to have it covered, that can be done."[18] To do it involved nifty plastic surgery, one that we instinctively believe only modern medicine could have known. Boys are more easily treated than men, he observed. Celsus's details described the two procedures: one is when there is a small defect of nature that is more easily corrected. The other is for "one who after the custom of certain races has been circumcised."[19] Jody F. Rubin has researched these two procedures as described by Celsus.[20] Since Rubin's study, Dirk Schultheiss, a physician in Hannover, Germany, joined some of his colleagues in reviewing the surgical procedures and carried this historical research down to Nazi Germany when some surgeons clandestinely performed essentially the same procedures as Celsus outlined for Jews' protection.[21] The first method is a cut of a patch of skin on the penis stem, loosening the epidermis and sliding forward the skin over the glans to reconstruct the prepuce. To be sure, scar tissue would form in the band, but the glans would be covered. The true epispasm (or uncircumcision) method is to incise at triangular cuts around the corona, the base of the glans, loosen the skin's epidermal attachment, and pull it over the glans. For each of these procedures two critical steps have to be taken. The skin drawn over the glans would not have epidermal exterior and interior layers, so a plaster separation must be made to allow a new epidermal layer to cover the prepuce. A similar separation would have to be made separating the new prepuce from contact with the glans.

Celsus describes this second procedure:

> If the glans has become so covered that it cannot be bared, a lesion which the Greeks call *phimosis*, it must be opened out, which is done as follows: underneath the foreskin is to be divided from its free margin in a straight line back as far as the frenum, and thus the skin above is retracted. But if this is not successful, either on account of the constriction or of the hardness of the skin, a triangular piece of foreskin is cut out from underneath, having its apex at the frenum [=a membranous fold of skin], and its base at the edge of the prepuce. Then lint dressing and other medicaments to induce healing are put on. But it is necessary that the patient should lie up until the wound heals, for walking rubs the wound and makes it foul.[22]

The second hurdle is that the man should have no erections for the weeks of post-surgical healing.[23] Celsus spoke of this impediment only by saying that in order to allay "erectile function," the patient should fast to reduce his libido. Now, this is too much to expect of virile, late adolescent male athletes. Besides, as we know from dieting, cheating is easily done as the days roll by into weeks. In the weeks of postsurgery, a single erection would undo the work. An eighteen year old, no matter how much he suppresses thinking about sex, will think about it anyway—he simply will. No citations in medical journals needed on this subject. To accomplish this period of quiescence, a plant was used to insure a successful operation. The surgical operation was directly linked with the Jewish Maccabean Rebellion (167 BCE) and the establishment of an independent Judea. Later more information will be related on the Rebellion, so now let us look at the plant that made this happen. It is the chaste tree, a small tree, whose name in Greek, *agnos,* meant "chaste" and whose name in modern scientific nomenclature is *Vitex agnus castus* L., meaning chaste tree. *Castus* in Latin means "pure" or "unstained," thus the tree is "pure chastity." Probably because the chaste tree's leaves resemble those of the willow, according to Pliny,[24] the plant's genus name in scientific nomenclature is *Vitex.*

The Chaste Tree Plant as Male Contraceptive and Sexual Suppressant

Priests ate of the chaste tree to prevent their having erections but it was in conjunction with other plants. The Pseudo-Galenic work *De remediis parabilibus*[25]:

> Things having power over stopping intercourse (*sunousiazeiv*): Priests (*ieraeis*) eat rocket (*Eruca sativa* L.), rue (*Ruta graveolens* L.), and the chaste tree (*agnon=Vitex agnus castus* L). Certainly with rocket seeds in conjunction with honey and fasting, one is not able to have an erection (*orthoisthai*).

Dioscorides and other classical writers make no such claim for rocket; he even said that rocket was an aphrodisiac, the opposite of the chaste tree.[26] Concerning rue, he said, "the seed quells the organ of generation when eaten and when drank."[27]

Before concentrating on the chaste tree, let us direct our attention to the plant rue that was also in Galen's preparation. No modern research in the databases indicates testing rocket for hormonal effects in mammals. Rue is different. From antiquity to modern times, rue is a well-known abortifacient,[28] but there is little mention of the plant acting as a male contraceptive or sexual suppressant. Published in 2007 was the result of a study testing the immobilization effect of rue on human sperm. The experimenters were physicians in the Department of Physiology, Medical Sciences, at the University of Tehran. They selected rue because of its use in traditional Iranian medicine as a male contraceptive. Inasmuch as the usual address for male antifertility measures is either condoms or vasectomy, a male equivalent of the female contraceptive drugs would be welcomed, presumably. When different doses of lyophilized aqueous extract of rue were placed in a dish together with live human sperms cells, immobilization was immediate, but the degree varied with the strength of the rue extract. After washing, the sperm returned to mobility in this in vitro experiment. Importantly the degree of activity was dose-dependent, with immobility occurring at a concentration of 100 mh/ml.[29] In 2005, two faculty members of the medical university in Jordan experimented using albino rats who ingested rue. Male rats had sperm motility and density in testicular ducts "significantly decrease[d]." These experimenters were not testing contraceptive uses but the aggression and sexual behavior in male rats. Conclusion: "Ingestion of *R. graveolens* markedly abolished aggressive behavior parameter in adult male treated rats."[30] These animal tests indicate an effect on male fertility for higher mammals. What we do not know is whether rocket seeds are the opposite of Viagra, namely that these seeds prevent erectile function. There is evidence, however, that the chaste tree has that effect.

Our attention turns to the chaste tree because of the persistence in traditional medicine and historical medical records for its effect on male sexual activity. Neither rocket nor rue has historical continuity in Western medical sources but the chaste tree does for suppression of sexual activity. Moreover, like pomegranates, mandrakes, and artemisia, the chaste tree is connected with ancient mythology. Science journals have many studies based on in vitro and various animal clinical trials that indicate that the chaste tree has definitive suppression of sexual activity and spermatogenesis in males. An Indian study employed male Wistar rats to test a related species, *Vitex negundo*, commonly found in India and selected (at least, that was inferred from the published account) because of its reputation as a male contraceptive in traditional Ayurvedic folk medicine. Extracts were obtained from seeds and in various concentrations. The rats were divided into five groups of four animals each and injected with varying concentrations once daily for fifteen days. They were killed and subjected to hematological and sperm counts analyses, including postmortem anatomical studies of various organs. Ranging by degrees according to concentrations, the spermatozoa count was significantly reduced progressively. The few remaining healthy sperm were likely "inadequate for achieving fertilization."[31] No behavioral changes were observed before death and postmortem examinations of organs revealed no organ abnormalities. The study concluded that "It is abundantly clear that the seed extracts of *V. negundo* have a profound effect on the sperm quality and the functioning of the accessory sex organs."[32] To test the chaste tree on female reproduction, experimenters found that mice given doses of 60 to120 mg on days four and six of pregnancy had 100 percent of pregnancies terminated through implantation disruption. For days eight to ten of pregnancy, 50 percent had their pregnancies terminated.[33] Consistently through the ages the chaste tree was used as an abortifacient, but its use as a male contraceptive is less well documented.[34]

Even though they are easy to obtain and control, rats are not always the best subjects for clinical reproduction studies. Another Indian study, published in France, tested the same species, *V. negundo*, on ten adult male dogs with a control group. Seed extracts were ingested every other day over forty-five days to test the influence on testicular and epididymal function. The plant "inhibits spermatogenesis as well as it alters epididymal function."[35] Moreover after withholding the treatment, spermatogenesis returned, thereby making the chaste tree a reversible male contraceptive.

Studies have tested the effects of the chaste tree on human males. Ingested the plant causes a prolactin-inhibiting action. In pharmacological terms it has a dopaminergic action based on stimulation of D_2-type dopamine receptors. A clinical test was conducted with twenty healthy males with variables depending on baseline values of the individuals.[36] The libido and fertility in males were drastically reduced. Importantly, there were no serious side effects, even with higher dosages, although it should be stated that the full pharmacokinetics and toxicology of chaste trees have yet to be defined.[37]

CHASTE TREE USES IN ANCIENT EGYPT AND ASSYRIA

Despite the near-certainty that the chaste tree grew in Palestine, the plant is not mentioned in the Bible; however, other peoples of West Asia and Egypt knew some of its medicinal qualities.[38] The ancient Egyptians knew the plant as *s'3m* (romanizing the hieroglyphs) but not for reproduction or sexuality suppression as seen in the surviving documents. If our identification of *s'3m* is correct, chaste tree [bark?] was used in bandages to reduce swelling, to help with constipation, and, its bark ground with water, to strengthen the teeth.[39] Assyrian medicine prescribed the use of the chaste tree. R. Campbell Thompson identifies two Assyrian words: ^{šam}Šišbânu (which could either be *Vitex negundo* L. or *V. agnus castus* L.) and ^{šam}Šana-a (as *V. agnus castus* L. or the chaste tree). The names appear in a list of small trees (which is what the chaste tree is). The first species (which Thompson considers ambiguous) was used on bandages similarly to the Egyptian use. The chaste tree alone was used on the feet for swelling and childbirth.[40] The childbirth use will be discussed later; the application on the feet is, again, similar to the Egyptian medical use.

THE GOD PRIAPUS AND PRIAPISM

An Assyrian medical recipe was used to treat priapism, a serious medical condition when a male maintains a continuous erection. The recipe does not have the pharmaceutical used to treat it, but the first part reads: "If a person continually has an erection when he tries to urinate...."[41] Another Assyrian source explains how serious priapism is: "If the soles of his feet are cold, the rims of his eyes are swollen, his penis is stiff and hard, (and) his navel is supple, 'hand' of curse; (if) he lingers, he will die."[42] The condition is a real one, a current affliction that is caused by neurological and vascular factors; just as the Assyrian text read, priapism is potentially fatal.

Priapus's mother, Aphrodite, disowned him because of her revulsion for his grotesquely large penis and his foul language. His paternity varied in the ancient sources ranging from Zeus, Pan, Dionysus, and Hermes; disowned by his mother, Priapus joined Pan and the satyrs. In one story, he attempted to have sexual liaison with the nymph Lotis while she slept, but the braying of an ass awoke her in time to escape. The gods and goddesses saved Lotis from his pursuit. They condemned Priapus with a perpetual erection from which he could derive no pleasure by intercourse or masturbation; with it came pain and isolation. Whereas Priapus was never a prestigious god, the abundance of statuettes and paintings (walls and vases) attests to his popularity among the common folk.[43] Without labels, Priapus is recognizable by—well, you know.

Galen, the physician in the Roman Empire, described priapism (the condition named after the god) and its remedy:

> Priapism is an increase in length as well as circumference of the entire male organ without desire for sexual intercourse, and without some acquired heat.... Some have outlined the condition as follows...a persistent, unchanging increase in size of the male organ or a persistent swelling. It obviously has a name derived from Priapus. For human beings sculpt as well as paint Priapus as one who by nature has such an organ...And to treat priapism, apply the things that have been discovered by experience to be naturally efficacious, namely give him [the patient suffering from priapism] yellow waterlily [nymphaia] to drink, the seed of the chaste tree and pale rue mixed with his food.[44]

About the yellow (or white) water lily, Dioscorides said that its root, ingested, allays wet dreams ("ejaculations during sleep"), and he added, "it effects a slackness of the genitalia for a few days if one kept on drinking it." Similar results are achieved with its seeds.[45] Unable to find modern science research on the possible effects of the white water lily on priapism, several Web sites say that it is an aphrodisiac, and it was employed in ancient Vedic medicine as a sedative.[46]

Chaste Tree in Classical Medicine as Male Sexual Suppressant and Contraceptive

The English name "chaste tree" has the same meaning as the Greek *agnos,* which means "chaste," "pure," and/or "holy." A similar meaning was conveyed in Latin. The name itself told only a small part of the story; certainly the plant was thought to suppress the male libido and was connected to the gods in a strange way. Aelian (third century CE) connected *agnos*' meaning with its suppressive effect on "sexual appetite/ *aphrodisiou kōluma*."[47] Among its uses was to suppress wet dreams or involuntary nocturnal emissions of semen.

Galen said that

> The fruit of the chaste tree...does not cause wind [flatulence] in the belly, both when unroasted, and even more so when roasted. And it keeps in check sexual impulses—both the roasted and the unroasted fruit do so—and the leaves and flowers of the shrub can do this same thing, so that it is trusted to effect chastity, not only when

> it is eaten and drunk but also when it is strewn underneath as bedding.... From all of which it is clear...that that chaste tree heats, then it dries at generate flatulent pneuma [air], it would have inflated the belly [womb?] and thus would have sharpened the sexual impulses. But... it not only does not spur them on but also by nature suppresses them.[48]

Pliny said that the chaste tree inhibits sexual desire.[49] Oribasius (fourth century CE) said that flakes of lead prevented wet dreams (*oneirotokos*) when placed under the loins just as do rue and "tender extremities of the chaste tree" in a person's bedding. He added that the same effect occurs when rue and the chaste seed are eaten.[50] Paul of Aegina repeated much of the same thing: "The Chaste-tree... consists of fine particles and dispels flatulence, whence it is believed to contribute to chastity, not only when drunk, but also when strewed under one [as in bedding]."[51] Galen was once a physician in a gladiatorial school, and he expressed his admiration for an athletic trainer who used chaste twigs in bedding to avoid wet dreams. Presumably Galen thought that prevention of nocturnal emissions helped his athletic prowess.[52] The use in bedding, as we shall see below, was also connected to religious ritual.

The Chaste Tree, Epispasm, and Circumcision

Dioscorides wrote that the chaste tree "slackens the organ of generation,"[53] a polite way of saying that the plant prevents or, at least, impedes an erection. As we saw earlier, Aelian wrote that the chaste tree (*agnos*) was so named because it was "an impediment to sexual impulse."[54] During the Middle Ages, the chaste tree was known as monk's pepper in English,[55] *Mönchspfeffer* in German.[56] The monks used the chaste plant instead of pepper to suppress sexual desires.[57] The monk's pepper name indicates a continuation of use given by the pseudo-Galenic work, which said that priests use the chaste tree to prevent erections. Dioscorides explained the association with pepper because, he said, the chaste tree seeds resemble pepper.[58] The Benedictine Rules mandated that monks were to sleep in separate beds and that "the younger brothers shall not have beds by themselves, but interspersed among those of the elder ones."[59] Monastic order took steps to curtail sexual activity. As late as 1620s Jean Prevost's medicinal guide recommended use of the chaste tree, as it extinguished a man's semen "to impede generation."[60] The natural question is: why, if this were truly a male contraceptive, did it not continue in use and have an impact in today's society? An answer or, at least, an address of the question is postponed to the next and last chapter.

Maccabean Revolt, 167–163

The evidence points to common-sense measures to curb male sexuality in classical and medieval societies. Let us return now to the Jews of the second century BCE. The circumstantial evidence is persuasive that the Jewish male youths employed under a physician's direction the chaste tree in order to have a successful postsurgical course of action in the epiplasm or circumcision-reversal operation. In 167 BCE Antiochus Epiphanus outlawed Jewish practices, including circumcision—the flash-point. The second policy as a means of Hellenization, celebration of drama in the theater, seemed to produce no emotional opposition. Measures against circumcision did! Mattathias, a common man who had had enough, killed a Hellenizing Jew who pushed for reform in his village; with his five sons, Mattathias took to the hills. Other Jews joined them and quickly an insurrection was underway. A year later Judah "Maccabeus" ("the Hammer"), Mattathias's son, achieved a major victory over the Seleucid (Greek) army. Subsequent events enabled an independent Jewish state by 163 BCE. These events, known as the Maccabean Revolt, turned on the action of a plant, the chaste tree along with some skillful surgery. Without the chaste tree, uncircumcision would not be possible. Its critical use in surgery was a factor in the events leading to the Maccabean revolt. James Michener captured the intensity of feeling on both sides, Greek and Jewish. The chaste tree's influence on history did not end with the Maccabean revolt.

The Romans' attitude was similar to that of the Greeks: circumcision was a loathsome assault on the body. Nevertheless, even after the Jewish-Roman War and the Destruction of the Temple (70 CE), the Romans did not legislate against circumcision probably in recognition that its enforcement would lead to public disorder. Although we do not have the wording of the legislation, under the Roman Emperor Hadrian (117–138) circumcision was forbidden on pain of death, not because of religious intolerance but because it was regarded as unacceptable, barbarous behavior.[61] The law, probably enacted in the year 132 CE, was not ethno-specific, and, even more to the point, it held as equally felonious castration and circumcision.[62] The Jews were not specifically targeted because it applied to a number of groups, including Egyptian priests. The prohibition of castration, a loathsome and public procedure, would apply to the religious rites in Magna Mater or Cybele worship. The next emperor, Antoninius Pius (138–161), reversed Hadrian's legislation in part, thereby permitting Jews to continue their practice. As quoted in the *Corpus Iuris Civilis' Digest* (and essentially in medieval law thereafter): "Jews are permitted to circumcise only their sons on the authority of a rescript of the Divine Pius; if anyone shall commit it on one who is not of the same religion, he shall suffer the punishment of a castrator."[63] In practice, the law forbade circumcisions of Jews' slaves and had the effect of preventing proselytizing. The fact that Celsus, the Roman medical writer, had the elaborated discussion of the procedures for circumcision reversal indicates that the surgical procedure was not historical antiquarianism; it was a current practice in his time, early first century.

Thesmophoria and Chaste Tree

Already in chapter two we saw the association and use of the pomegranate in the ancient women's ritual known as the Thesmophoria, the annual cult of Demeter in which Greek women go on a retreat for a period of three to ten days in the late fall. Although the festival was for fertility, on the second day the women—for only they were present—fasted for the day and slept that night on a bed of chaste twigs and leaves. Pliny wrote: "The Greeks call it *lygos* [an alternate name], sometimes *agnos*, because the Athenian matrons, preserving their chastity at the Thesmophoria, strew their beds with its leaves."[64] Writing probably a few years earlier than Pliny, Dioscorides related the same information as Pliny about the chaste tree and Thesmophoria but added "because the women make it a point of religion."[65] Heinrich von Staden's study of the Thesmophoria and the chaste tree confirms that the symbolism for the day and night's activities was for abstinence from food, comfort, and sexual activity. By the women sleeping on a bed of the chaste tree (twigs, leaves, bark), von Staden saw its meaning as "not only a regression into precultural wildness, but also a new beginning, at the very inception both of culture and of human life."[66] Above, we saw that Galen spoke of chaste tree bedding as a way for males to suppress their sexual desires whereas Oribasius and Paul were not gender-specific when they asserted that chaste tree bedding promoted chastity. To us moderns it is incongruous that married Greek women would be preserving chastity, but the ancients saw chastity not so much as virginal as pure in actions. Von Staden saw a paradox in that a festival for promoting fertility would have as a component a "suppressant of sexual desire."[67] The practice was more than symbolism because the Thesmophorians believed that their ritualized actions either suppressed sexuality or were instructive about the same. If merely symbolism for a back-to-nature rite, then a set of twigs from any bush or tree would be equally serviceable. Paul of Aegina said that products of the chaste tree either eaten or in bedding promoted chastity, and Galen said in bedding wet dreams were prevented. Our modern science gives evidences that the chaste tree was meaningfully named: it suppresses sexual activity in men. We cannot infer, one way or the other, whether transdermal absorption was sufficient for a pharmacological action when a person slept on the chaste-tree bedding. Certainly the action at the Thesmophoria was partly, not exclusively, symbolic. Von Staden assembled many more references in the works of Homer, the Homeric hymns and other pieces of Greek literature for one to conclude safely that sleeping on chaste-tree bedding was rooted in the popular culture.

I cannot restrain a speculative indulgence: in going to the Thesmophoria, women were leaving their husbands alone for days. Surely some women left some chaste tree ground bark or seeds in the leftover stew so that their men would not stray from home in lustful quest. The festival would be instructive as well as religious. Women would want to know both how to promote sexual activity and how to suppress it: in a woman's life, there was a time to have babies and a time when not to become pregnant. They knew both male and female contraception.

THE FEMALE HERB

In 1571, Gerard, writing a famous herbal, allegedly called the chaste tree the "female herb."[68] Heretofore our focus was on the plant's effects on male reproduction in ritual, myth, and medicine. The banner headline is males' use of the chaste tree (wittingly or unwittingly), but it probably affected more women than men historically. Gerard wrote: "The decoction of the herbe and seede is good against paine and inflammations about the matrix, if women be caused to sit and bathe their privy parts therein; the seed being drunk with Pennyroyal bringethe downe the menses."[69]

In 2003 Christopher Hobbs issued the second edition of his small, well-documented book entitled: *Vitex: The Women's Herb.*[70] In a brief excursus in the plant's history, Hobbs said that the chaste tree's popularity waned in the early eighteenth century; the return of medicinal uses for the tree came after the researches of Gerhard Madaus in the 1930s in Germany. Madaus patented an extract made from the dried fruit berries that he called Agnolyt®. Most clinical and laboratory studies of the tree's actions employed the Agnolyt preparation. Using Agnolyt, several studies in the 1950s found that the chaste tree stimulates lactation.[71] These studies confirmed what the writer of the Hippocratic work *On the Diseases of Women* wrote approximately 2,400 to 2,300 years ago when the author singled out chaste tree fruit for the lactation action.[72]

Modern researchers find that in a three-month therapeutic period the chaste tree has a "strong rate of success" regarding various menstrual bleeding disorders: amenorrhea, hypermenorrhea, and oligomenorrhea. The German Commission "E" approved the use of chaste-tree berries for menstrual irregularities, PMS, and mastodynia (neuralgia of the breast).[73] Significantly, the studies are of human subjects. Some thirty-two clinical studies between 1943 and 1997 attest to the chaste tree's medicinal qualities.[74] One double-blinded study showed significant effects of the chaste tree. Women having taken the chaste plant showed an "inhibitory to pituitary prolactin and therefore beneficial for the treatment of premenstrual mastodynia" when compared to those on a placebo.[75] Mastodynon® N is a German proprietary drug containing a preparation of the chaste tree. In 1998 Mastodynon was administered to ninety-six women (36 of whom suffered from secondary amenorrhoea; 37 with luteal insufficiency; 27 with idiopathic infertility). Women treated over a three-month period showed improvement with amenorrhoea or luteal insufficiency, and pregnancy occurred at twice the rate of the placebo group.[76]

The chaste-tree plant is especially effective for PMS linked to abnormally high estrogen levels. Scientists hypothesize that the plant acts on the pituitary gland for chemical signals to balance female sex hormones.[77] Heinrich von Staden's study of the chaste plant's use in the works attributed to Hippocrates is in line with modern studies. A number of prescriptions are given for various womb disorders and menstrual problems, especially those associated with vaginal discharges (bleeding). Three examples from *Diseases of Women II*: "If blood flows violently from the uterus, leaves of chaste tree with dark wine. . . ."; "If the lower abdomen aches because of a watery flux, grind up chaste tree in honey and apply it nicely as a poultice"; "When worms occur in a woman's genitals . . . a mixture is made of the fruit or leaves or the chaste tree." The last prescription was for a restoration of menstrual regularity to promote fertility where we assume that "worms" may be interpreted by us as meaning a disorder.[78]

Dioscorides wrote,

> It [chaste tree] draws down milk and it brings on the menses when a quantity of one *drachma* is drunk with wine; but it also slackens the organ of generation, and it affects the head inducing deep sleep. The decoction of the plant itself and of its fruit helps in sitz baths for conditions and inflammations of the uterus.[79]

In Dioscorides and the Hippocratic works, the means of address varied from ingestion, vaginal suppositories, and epidermal application.

Soranus, the writer in Greek on gynecology, spoke about the chaste tree's effects on male and female bodies with, as expected, more attention given to women. Soranus was writing about a woman's disease whose symptom is the "flux of semen [seed] ... in women." In antiquity there was a debate over whether women produced "seed." Aristotle, among others, argued that only men produced seed and, when deposited in the womb, could be accepted or rejected by a woman's body. If accepted, the woman' body supplied nourishment to produce the offspring but not seed (which would make her more of an equal partner). Other ancients, including some of the Hippocratic writers and Soranus, believed that women's bodies produced seed necessary for conception, thereby subtly elevating women's status. Exactly what flux of the woman's seed meant in corresponding modern medicine's nosonomy escapes my ability. Soranus was guided by the Methodist School of medicine, which classified symptoms under such terms as contraction, restriction, wasting, condensing, and loosening. In part, he describes this disease and its therapy as

> The flux of semen [*gonorrhoia*] occurs not only in men, but in women too. It is a discharge of seed [*sperma*] without desire or erection [or "tension"] and happens at short intervals, so that the body become pale, loses strength, and is consumed. For little by little the substance [*hyle*] of the body flows into the uterus, undergoing a slight change in the morbid organ, as do the tears of people suffering from ophthalmia [inflammation in the eye]. The disease is usually chronic and belongs to the "lax" kind. [various physical therapies and drugs are described] ... One should give [her] as a potion one *drachma* [=approx. 3.4 g.] of the root of winter-cherry [*Physalis alkekengi* L.], dried in the shade, together with water, or the seed of the chaste tree in like manner or the seed of hemp or rue.... One should also avoid all sexual stimulants; thus one should not show her paintings of shapely form nor tell erotic stories.[80]

A man's discharge of seed, as stated by Soranus, is likely the suppression of wet dreams as related earlier. Soranus is alleging that the chaste tree and winter-cherry should be taken in addition to a woman's avoidance of sexual stimulation. Dioscorides recommended winter-cherry as good for erysipelas, shingles, lachrymal fistulas, and severe fluxes.[81] The Hippocratic author of *Women's Diseases*, Dioscorides, and others prescribed the chaste plant to stimulate menstruation (Gerard's "bringethe downe the menses").[82] This action was commonly meant also to cause an abortion if pregnancy were the cause of menstrual interruption. That said, the chaste-tree plant was not prominently mentioned as an abortifacient. Its action historically was more corrective of menstrual problems, just as we currently employ it.

HERMES, CHASTE TREE, AND ANIMAL HUSBANDRY

In Greek myth, Hermes was the son of Zeus and the shy, nymph Maia, who lived in a cave where she bore their son. Not an ordinary son, Hermes was called a "robber, a cattle driver, a bringer of dreams, a watcher by night, a thief at the names, one who was soon to show forth wonderful deeds among the deathless gods."[83] Hermes's deeds began during the night of his first day after birth; he rose "cunningly" from his crib, miraculously, we would say, and stole some fifty cattle from Apollo's herd. Cleverly he had their hoofs pointing in opposite direction from which he drove them. Upon returning to his cradle in the early morning, his mother had discovered his absence: "You rogue," she said and told him to report to his father, Zeus. An old shepherd saw the baby Hermes driving "strong" cattle "through many shadowy-mountains and echoing gorges, and flowery plains."[84] Discovering his cattle missing, Apollo received the testimony of the old shepherd and confronted Hermes, who denied that he was a thief, seeing as how he was only a one-day-old child. At Hermes's urging, Apollo took the matter to Zeus who heard the evidence. Zeus asked Hermes to take them to where the "strong" cattle were hidden. Seeing his cattle, Apollo feared that, if Hermes had the strength to drive cattle on his first day of life, he would be a danger as he aged; Apollo tied him with chaste-tree binding. The bands would not hold him but fell to the ground and began growing to "cover all the wild-roving cattle."[85] Apollo's astonishment was tempered by delight when Hermes handed him a lyre with which Apollo instinctively produced song, the lyre becoming Apollo's symbol. Gladly did Apollo exchange his cattle for the lyre. Hermes said, "But I, o Far-Worker, will graze the filed-dwelling cattle on the pastures of the mountain and on the horse-feeding plain where the cows, mounted by the bulls, will give birth abundantly in both females and males."[86] The Greeks and Romans regarded Hermes as the father of animal husbandry, responsible for livestock reproduction, "the keeper of cattle."[87]

Other classical sources provide abundant evidence that Hermes was the father of animal husbandry. For examples,

> She [Hekate] is good in the byre with Hermes to increase the stock. The droves of kine and wild herds of goats and flocks of fleecy sheep.
>
> —Hesiod, *Theogony*. 443–445 (Evelyn-White trans. p. 111).

> Ilioneus the son of Phorbas the rich in sheep flocks, whom Hermes loved above all the Trojans and gave him wealth [by multiplying his flocks].
>
> —Homer. *Iliad*. 14.491 (Murray trans. 2:103).

> I also pray Hermes, the god of the shepherds, and Pan.
>
> —Aristophanes, *Thesmophoriazusae*. 970.[88]

> Hermes is the god who is thought most to care for and to increase flocks.
>
> —Pausanius. *Guide to Greece*. 2. 3. 4 (W.H. J. Jones, trans. 1: 261).

As we have found ancient myths are built on a kernel of truth. Hermes did not steal the god's cattle, but he separated them, bound them with the chaste tree, and multiplied their reproduction. Breeding is the core to animal husbandry, by both increasing fertility and selection of animals for breeding. Learning from the story, Hermes knew the secret of the chaste tree, and he applied it. He may have fed it to cows and other female livestock to increase their fertility, since we have modern evidence that it aids menstrual problems and increases fertility. He may have fed it to bulls and male livestock to thwart the fertility of those animals that he did not wish to have progeny. How this skill was applied escapes notice in the meager records about ancient rural life. Reasonably, it is too much of a coincidence not to recognize the tie between the founder of animal husbandry and the chaste plant. Hermes, the god, taught humans how to care for animals by breeding. Such a lesson is significant in human history, although underappreciated by historians.

Epilogue on the Chaste Tree and Prologue for Chapter Six

Some years ago when first I became intrigued with the chaste tree's actions, I gave a lecture to a botany group at a small and excellent Baptist women's college. Although the chaste tree accounted for only a small part of the lecture, when the lecture concluded a number of students gathered around the podium, as one student asked, and others nodded in approval: "Where can we find this chaste tree?" Thinking that underlying their question was the wish to control the "gentlemen callers" by actions opposite to those of Viagra, I equivocated and replied, correctly, that it grew in the State arboretum. Its bark and leaves were safe in the garden, I thought. Even though sufficient modern science studies support the plant's action as employed in antiquity, its use as a pharmaceutical (or nutritional supplement) needs more human clinical studies before one can advocate its revival. The college women were not told that the small tree also, as it happened, grew beside the front door of their science building where the lecture was held. The knowledge about the chaste tree was once a part of popular culture and its stories embedded in their religion. Why is it that the modern, well-educated women in botany were unaware of its historical and pharmaceutical legacy? The answer of a lost heritage will be addressed in the final chapter. Also, we shall have Hermes reappear, reborn as a god with powers dwarfing his cattle-stealing deeds, phenomenal though they were, when we consider that he was at the time only one day old. Hermes learned quickly the chaste tree's powers.

CHAPTER 6

HERMES, HERBS, ELIXIRS, AND WITCHES

> Truly, whoever knows how to do these things [alchemical preparation] would have the perfect medicine (*medicinam),* which the philosophers called the elixir.
>
> —Roger Bacon, *Opus Minor,* 314.

Hermes, the young god who rounded up Apollo's cattle on his very first day of life, was destined to have a long life, longer and more renowned than most pagan gods or goddesses, even Inanna. Hermes went to Egypt and there morphed into a secret-knowing, slow-releasing-of-information god: he acquired knowledge, a prerequisite many theologians of antiquity would say for divinity. Hermes's knowledge just was not omniscient knowledge but facts on how to do things. Practical knowledge was a diversion of the intellect according to Plato and many tenets of classical philosophy. In Greek, Hieroglyphs, Hieratic, Latin, Arabic, and the vernacular languages of the Middle Ages, the secrets of *Hermes Trimegistus* ("Hermes Thrice Great") stimulated a line of thought that combined with the Chinese concept of an elixir to produce a distinctively Western attitude about drugs. The new way of thinking was substantially different from that of the classical Greeks and Romans. The powers inherent in the pomegranate, mandrake, artemisia, and chaste-tree plants turned toward a new direction. The course was diverted, because its association with magic caused them to appear as black or evil magic. The herbs sacred to Inanna, Aphrodite, Demeter, Artemis, and Hermes came to be associated with witches and nefarious forces and not with goddesses (and, in Hermes's case, a god). Remember: the first witch was Lilith, who made a home in the trunk of the pomegranate tree, thereby robbing Inanna of her power. The secret knowledge had to be suppressed and, largely, it was. This chapter explores the three forces that turned the herbs of goddesses into those of the devil. Simultaneously and ironically developments over a long period of time gave us the hope for miraculous cures, divorced from pagan gods or the God of Jews, Christians, and Muslims. The three streams of developments were Hermetic literature, the concept of the elixir, and witchcraft.

THRICE-GREAT HERMES

Treatises ascribed to Hermes are often referred to as the *Corpus Hermetica*, but "corpus" or "body" is applied to those works only from the classical period. Unlike the Hippocratic Corpus, most of which was written in a little more than a century, the Hermetic works span many centuries ranging from antiquity through the sixteenth century. Conventionally, however, most scholars refer to the *Corpus Hermetica* as those works written in antiquity until the late Roman period.[1] The anonymous and pseudo-anonymous authors of these works had diverse backgrounds and interests, and wrote in various languages. Nevertheless, like the Hippocratic works, there are a few characteristics that underlie virtually all of the works: they have no identifiable religious home, excepting perhaps in obscured ancient Egyptian religion. Pagan, Jew, Christian, and Muslim alike considered the works their own; the writings transcend other religions but have traces of Egyptian and Greek thought with an element of practicality built on philosophical underpinnings; they profess to be "secrets" whose utility must be guarded by their readers, the restricted few. The contents include herbs, stones, animals, sigils (magical signs), alchemy, and astronomy/astrology.

In the Middle Ages, people were clearer about the origins of the Hermetic writings than ancient people who mixed classical and Egyptian sources. According to medieval beliefs, Hermes was the messenger of the gods; he was the conveyer of information. He went to Egypt, married the daughter of the Pharaoh, learned the secrets of the Egyptians, and deposited them in the library, presumably the one in Alexandria. When the library burned, the secrets of Hermes were lost except for a few pieces of text that somehow survived. During the first millennium of our era and through the sixteenth century, these treatises circulated because of their popularity. One who discovers a part of the secrets may divulge the secret therein to a few. Those few must keep the information confidential lest by becoming known to common people (vulgarized), the secrets lose their power.

ANCIENT HERMETICA

The early Hermetic works were in circulation by the first century of our era; from the third century BCE, Hermes was said to be their author. Scholarly study detects that the early treatises' contents were derived partly from ancient Greek works but predominantly from ancient Egyptian priestly lore. Hermes and his Egyptian counterpart, Thoth, melded into one.[2] Thoth was the tongue and heart of Ra, the principal Egyptian god, and, as such, stood beside him and was his spokesman, similar to a messenger. Thoth was responsible for writing, magic, the mathematical and grammatical sciences, and was the judge of the dead. He kept the list of each person's good and evil deeds. Hermes succeeded Thoth, even though the works attributed to Hermes are more Egyptian than Greek, in a way reminiscent of a paraphrase from Horace: "Captive Egypt captured its captive."[3] The Egyptian priesthoods studied geometry, arithmetic, astronomy, and medicine. Two kinds of knowledge are accessible to humans: *epistēmē* (concrete, specific data, sometimes too loosely translated as "science") and *gnosis* (knowledge derived from *logos* or pure reasoning supplemented by understanding related to faith but not its synonym).[4] Humans approach God (or the gods) through knowledge. Such Gnostic sentiments were not alien in Hebrew thought. "Wisdom found no place where she could dwell, and her dwelling was in heaven;" so read 1 Enoch 42: 1 (Pseudepipgrapha); Proverbs 3: 13–14: "Happy is the man who finds wisdom, and the man who gets understanding, for gain from it is better than gain from silver and its profits better than gold."

Gnostic ideas strongly pervaded Hermetic philosophy: a virtuous person moved from ignorance to *gnosis* (revealed knowledge) but encountered a danger, since some knowledge could lead to evil purpose.[5] Virtuous people seeking knowledge had to apply it in a correct, beneficial way. One Greek Hermetic text reads in translation: "The vice of the soul is lack of knowledge (*gnosis*). A soul that has gained no knowledge of the things that are, and has not come to know their nature, nor to know the Good, is but blind.... On the other hand, the virtue of the soul is knowledge. He who has got knowledge is good and pious; he is already divine."[6] Knowledge is delivered indirectly, as Hermes explains in a treatise known as *Aesclepius* (13): "Many make philosophy hard to understand, and corrupt it with manifold speculations... in this way, by a cunning sort of study, in which philosophy will be mixed with diverse and unintelligible sciences, such as arithmetic, music, and geometry."[7] To us moderns, any assertion is absurd that arithmetic and geometry are irrational by their nature. We must look deeper into the ancients's souls. "Earth is irrational; heaven is rational,"[8] a work of Hermes explains. The facts (*epistēmē*) are known through the sciences, but facts enable human reasoning only through divine revelation.[9] This knowledge is revealed in stages, not at once; when known, knowledge is power (*dynamis*), the very word used to describe the power of herbs.[10] The danger lies in how this knowledge is used: for good or for evil?

The direction in which these facts point is toward magic—white or black, good or bad magic. Pliny the Elder said it well:

> The most fraudulent of arts [magic] has held complete sway throughout the world for many ages. Nobody should be surprised at the greatness of its influence, since alone of the arts it has embraced three others that hold supreme domination over the human mind, and made them subject to itself alone. Nobody will doubt that it first arose from medicine, and that professing to promote health it insidiously advanced under the disguise of a higher and holier system... [then] it added the powers of religion.

As Pliny implied, medical knowledge was very much a central part of Hermetic works. Clement of Alexandria (d. 215), a Christian father, explained that the books contain "all the wisdom of the Egyptians, [including] treating the constitution of the body, maladies, organs, remedies, the eyes, and questions related to women."[11]

The power of Hermetic philosophy (called, by Pliny, magic) is through specific knowledge of things bringing us to the divine, not by reason but by revelation. To the classical Greeks, truth was the product of free inquiry; to late antiquity and early medieval peoples, truth was known through the scriptures and revelation. Written to an unspecified king, one Hermetic work dismissed Greek philosophy, but ironically, the author used the Greek language for his dismissal:

> The secrets so holy may not be revealed to Greeks, and that the Greek mode of speech, with its [*lacuna:* "arrogance" or "disdainfulness"?] and feebleness, and showy strength of language and cogent force of the words. For the speech of the Greeks, my King, is

> devoid of power to convince; and the Greek philosophy is nothing but a noise of talk. But *our* [wisdom of Hermes] speech is not mere talk; it is an utterance replete with workings.[12]

A modern scholar, Garth Fowden, wrote: "The magician's potential power [from knowing the secrets] was considered to be unlimited, certainly equivalent to that of the gods, . . . the corollary of the Egyptians' belief in the dynamism of words and spells."[13] Some Hermetic works contain specific and detailed information whereby one may learn nature's secrets, which in turn can be harnessed for application. One Hermetic work disparaged the Greeks as engaging in "inane foolsophy of speeches [while] we [possessors of wisdom], by contrast, use not speeches but sounds that are full of action."[14] Thus, it is through knowledge of "things" and "action" that we approach God. In explaining this, Stobaeus, a Byzantine Christian of the fifth century, said: He who knows these things "is able to form an exact conception of God; nay, if I am able to speak boldly, he is able to see God with his own eyes, and having seen God, to be blest."[15] Artapan, a Jewish writer in Greek (early second century BCE), explained the practical knowledge as coming from the Egyptians through a combined name of Thoth-Hermes: "[Thoth-Hermes] taught the Egyptians [about] navigation, the lifting of stones with cranes, weapons, water pumps, war machines, and philosophy."[16] Much of the Hermetica, especially those treatises known and perhaps written in the medieval period, is about practical subjects, the secrets of nature, but the information is related vaguest and a certain amount of purposeful obscurity. Explicit acknowledgment of the obscurity was outlined in a letter in Greek from Asclepius (Hermes' friend) to King Ammon:

> My teacher, Hermes—often speaking to me in private, sometimes in the presence of Tat—used to say that those reading my books would find their organization very simple and clear when, on the contrary, it is unclear and keeps the meaning of the words concealed; furthermore, it will be entirely unclear (he said) when the Greek eventually desire to translate our language to their own and thus produce in writing the greatest distortion and unclarity.[17]

The Hermetic works were not "Herbal Medicine for Dummies." While trying to understand a Hermetic work on herbs, minerals, or alchemy in medicine, I liken Hermetica authors to writers of computer manuals. The authors really do not want me to understand everything (or else I would know what they knew), and so they deliberately leave out a step or two, perhaps assuming that I might know it intuitively. Alas, I am not the equal of a computer programmer.

On the Fifteen Stars, Fifteen Stones, Fifteen Herbs, and Fifteen Sigils

In researching Hermetic works, I found in European manuscript collections a number of manuscripts attributed to Hermes on the fifteen stars, stones, herbs, and sigils (or engraved images). Each star has a secret the author's treatise will disclose, although not explicitly. Similarly, each star has a specific herb and stone, which the star has impressed on the plant or mineral. Each has a secret, a power which, when known, will benefit its reader. The treatises or sections of treatises on herbs have the familiar medicinal uses, many of which are to us rational, but also each herb can produce effects that we would classify as magic. Similarly with stones: a diamond is "indomitable" and cannot be broken and, for this reason, it is the wedding stone; jasper prevents conception by keeping its wearer from licentiousness; sapphire overcomes envy; chalcedony causes its bearer to win causes; and emerald, when worn, makes one's words persuasive. Most of the stones' powers fall under what we would call psychological. Some powers are magic, such as the emerald, which also permits one to foretell the future. The proper stone can control almost any emotion; one stone (whose name will be purposefully omitted) causes a speaker (as in a classroom) to stop talking, a secret that I shall keep to myself. Each herb and stone has a corresponding astrological sign or constellation that impresses its virtues on earthly artifacts. Finally, each herb, stone, and star corresponds to a sign engraved on stone. Initially in looking at these images, I thought them merely scribal scratches, because I could detect neither letter nor animate object. In time, it occurred to me that the signs appear to be a medieval scribe's attempt to reproduce cuneiform, except that the characters are reversed, as on a signet ring, to produce a cuneiform word.[18] Although predominantly ancient and medieval authors of Hermetica assigned the credit for the ancient knowledge as coming from the Egyptians, nevertheless, the "Chaldeans" or ancient Mesopotamians are credited with some of the "secrets."[19] At least some of the information imbedded in this treatise appears to have come from the Mesopotamians.

IMPORTANCE OF HERMETICA

God has placed in each plant, stone, and star a secret that, when discovered, will be of use to humans. "God will come to meet you," Hermes stated.[20] God, all-powerful to be sure, was virtually synonymous with matter or nature. "This material God, the Kosmos," Tat said to Hermes.[21] And, in a discourse addressed to Hermes: "All things are in God."[22] In a prayer: "O holy Knowledge [*dynameis*, pl. used], by thee am I illumined."[23] God has placed in humans the potentiality for *dynamis* [s.], perhaps better translated as "power" or, even better yet, the "potentiality for power."[24] The Hymn concludes with *enegeia tōn dynameōn <mou>*, or, when one's *dynameis* (powers of knowledge) operate, it is God who operates them.[25] "For by the working of the mind, you have come to know yourself and our Father."[26] One should not conclude that this is pure pantheism, because it is not. The cosmos had a creator: God, who was greater than his creation. God gave to humans the power of knowledge if they were smart enough to learn, understand, and use it. Earlier historians championed Sir Francis Bacon as a founder of the Scientific Revolution when he proclaimed: "Knowledge is power/ *Scientia potentia est*." His proclamation was not new. At least a millennium earlier than Bacon, the Hermetic works had asserted that knowledge was power. The essential feature is that knowledge was for utility, the benefactor of humankind.

The period of late antiquity and the Middle Ages regarded the knowledge of medicinal herbs as a power given to them by God, but the herbs' secrets had to be discovered. All things have secrets that, when known, are of benefit to humans.

Herbal knowledge, as all knowledge, could benefit the knower both physically and spiritually. This view of knowledge was in sharp contrast to classical philosophy. Greek philosophers (e.g., Heraclitus, Plato, and Aristotle) sought knowledge (*logos*, or "pure reason") in order to grasp the Absolute and to understand themselves through an inquiring mind ("Know Thyself"), and never to better their material being. Fired by the existence of Hermetic secrets, the medieval mind sought knowledge, but, in a basic departure from the ancient mind, medieval peoples blurred the distinction between science ("natural philosophy," to them) and magic. Magic was a means of getting nature to act unnaturally by a suspension of regular laws. Hermes "Thrice-Great" proclaimed that, through magic, one could obtain wisdom, which is the ability to perform miracles. In 1493 Germain de Ganay received a reply from Hermes himself in answer to some questions. Hermes said,

> Study generates knowledge; knowledge bears love; love, likeness; likeness, communion; communion, virtue; virtue, dignity; dignity, power; and power performs the miracle. This is the unique path to the goal of magic perfection, divine as well as natural, from which all superstitious and diabolical wizardry is totally separated and confounded.[27]

God was approached through magic, not superstition, because superstition comes from the devil. In bits and pieces, the medieval mind wanted to know practical details of how to manipulate nature, not conform to it. The Hippocratic physician wanted to learn the ways of nature so that he could help nature perform the healing process. By the thirteenth century, the medieval medical personnel no longer referred to themselves as *medici*, because they were *physici,* or, in English, "physicians." They sought to be "controllers of nature," and not the Hippocratic assistant or enabler of nature. Nothing can be a more optimistic proclamation than that learning practical detail made life easier for the virtuous people, for example, how to plow better a field, to make a better loom, or to discover the good and bad power of Artemisia. Once learned, these practical "secrets" were propelling virtuous people in the direction of God. No wonder this optimism drove away the shades of ancient gods. A great gap opened between the medieval mind and the ancient mind: the ancients were informed of the herbs sacred to the goddesses who bestowed benefits, if their rituals were performed and acknowledgment made of the gifts from the divine. Medieval peoples were given the potentiality of knowledge by God; they could on their own merits discover the medicinal secrets, since that was God's will. God placed those secrets within each herb and stone so that they might be discovered.

ELIXIRS

The concept of elixirs came from the ancient Chinese, not from the ancient Egyptians or the Mesopotamians. In China, as in the Hermetic works, there is a close connection between alchemy and medicine. The big difference was that in China the concept that special substances could bring immortality or, at least, prolonged youth. There was the "external elixir" known as *wai tan* that controlled vital substances with emphases on metals and drugs. The corresponding elixir concept was *nei tan* or the "inner elixir," which engaged the body physically and mentally through its energies (e.g., seminal essence, meditation, and breathing).[28] Through the outward and inner spirit a person could achieve an elixir that could prolong youth for decades, some said a thousand years, or, according to other Chinese accounts, material immortality. The process was purification of the external and/or inner elixirs. The concept evolved from Chinese philosophy and religion, but over time, emphasis turned from the mental aspects ("a comprehensive regimen of mental and physical disciplines")[29] to the actual process of how to discover the external elixir itself. As conceptualized in Islamic and Christian worlds of the medieval periods, the key was the so-called Philosopher's stone, which was not a stone as such but resembled a stone insofar as fire could not touch it.[30]

Fire was what caused transmutation; the elixir was (to be?) a product of the stove's transformative powers. Some Chinese regarded a goddess of the Stove as the spiritual guide; she was pictured as a beautiful old woman in red dress, her hair knotted on top of her head. She was the inspiration for alchemy. The Emperor Süan (*ca.* 60 BCE) was intrigued by the prospect of obtaining the (or an) elixir for immortality: to do this, he made sacrifices to the five sacred mountains, the last one of which was medicine. Intrigue turned to greed when the emperor was informed that the magicians (alchemists, as it were), who could transmute substances and obtain the elixir, could also make gold. The concept of such a possibility as immortality, youth, and abundant riches with gold traveled widely and for an extended period. In American history we know of its legacy through, for example, Ponce de Leon's search for the Fountain of Youth in what is now St. Augustine, Florida. Between Florida and the Chinese elixir much happened to transform the way we think about drugs.

George Sarton, the founder of the history of science in the United States, said that historians should exclude magic and superstition from their researches because they concern nonreason, and science was about reason.[31] Historians today reject Sarton's methodology. Whereas today rational science considers the elixir of life as magic and superstition at worse, fanciful and extravagant at best, the concept influenced early medical science immeasurably.

From China the concept of the "elixir of life," known in Latin as *elixir vitae*, captured the imagination of the Indians. In the Tantric Period (1300–1550 CE) of Indian medicine, the medical works of the Charaka and Susruta were filled with recipes, predominantly herbal, many of which we might judge "rational." Over decades extending to centuries, preparations appeared that were the products of the alchemists's ovens and vials. They included recipes that contained mercury, iron, copper, and other metals. Acharya Rây, a scholar of ancient India's alchemy, said of these recipes: "Although they could [allegedly] secure immortality [and] revive the dead, [they] were found to be helpful accessories in medicine."[32]

Islamic cultures took the elixir concept to a higher level of attention or, better, infatuation. First, the word *elixir* is of Arabic derivation from "al," as a definite article, before *iksīr*. The etymology may be an Arabic corruption from the late Greek *xērion*, a desiccative powder possibly at first to treat wounds, but its etymology is uncertain. The Arabs adopted the word elixir from the Chinese concept into a definition that revolved around two meanings: (1) a preparation used to change metals into gold and virtually interchangeable with philosopher's stone; (2) a medicine or drug capable of prolonging life, or even preventing death. The double meaning was, directly or indirectly, related to the bifurcated Chinese concept.

As Hippocrates's name was applied to a corpus or body of medical writings, so also was a corpus of more than 2,000 treatises attributed to Abu Musa Jabir ibn Hayyan (*ca.* 721–*ca.* 815) that circulated in Latin translations under the name of Geber, for Jabir. So far the task of separating the genuine Jabir from the pseudo-Jabir has escaped our scholarship. Although some attention was given to gold manufacture and the medicinal elixir, most Jabirean works were practical, such as how to make glazes, tannins, dyes, and waterproofing for cloth.

Alchemy was embraced and celebrated in the Latin West and Greek East from information derived both from translated works from Arabic and ancient Greek but also later Roman and medieval alchemical writers galore in Latin and various vernacular languages. The strongest, most eloquent spokesperson for the new science—for science it was called—was Roger Bacon (d. 1294), a Franciscan. He wrote: "Truly, whoever knows how to do these things [alchemical preparation] would have the perfect medicine (*medicinam*), which the philosophers called the elixir."[33] Medicines "have no power without fermentation.... This is a secret of secrets of which the generality of physicians is completely ignorant."[34] Proper (meaning effective) medicines are not simply the sum of their parts (herbs, minerals, etc.) but need to undergo a fermentation process, according to Bacon. The necessity for medical study is to master alchemy, astronomy, and *agricultura philosophica*. The latter is achieved by specific or particular knowledge of things gained "by way of experience"; it is "the power of recognizing the simple medicines derived from plants and animals."[35] These three sciences certify all conclusions of natural philosophy *per experimentiam*.[36] In viewing Bacon's remarks from our distance in time, it would seem to be strikingly modern as when he spoke of *scientia experimentalis*,[37] but this conclusion is not merited. By embracing the mysteries of the elixir and alchemy, Bacon and numerous other natural philosophers through the sixteenth century moved the pharmaceutical empirical science of antiquity to the realm of mysterious quasi-magic, all in the name of "science" as they understood science to be.

Living at the same time as Bacon and dying a year later than Bacon, a Bolognese physician named Taddeo Alderotti (d. 1295) was enthusiastic about the new medicines made from distilled wine (*aqua vitae*) mixed with various herbs which were "of inestimable glory, the mother and mistress of all medicine."[38] Muslim alchemists invented distillation and thereby made a *spiritus* advance (pun intended) throughout Europeans' gullets and into their medicine cabinets as well. These "wondrous" medicines, Alderotti proclaimed, were useful for melancholy ("makes one happy"), weak eyesight, epilepsy, paralysis, deafness, and cleaning wonders.[39]

Elixir and Garden of Eden

In Christian beliefs, the elixir concept was connected to the Garden of Eden and the Fall of Humans. The serpent, the "most subtle of creatures," told Eve and, through her, Adam, that if she ate of the Tree of Knowledge, the two of them would not die and would be, thereby, immortal.[40] The tradition became—because it was not implicit in the Genesis account—that the serpent was the instrument of the devil or evil personified. Why had the devil so acted?: because he, feeling himself inferior to God, wanted to alienate God from his created beings, Adam and Eve.[41] Paracelsus saw alchemy as the means of extending long life and, even though he is regarded as one of founders of modern science, Paracelsus spoke of "magic" and elixirs. "We see the age of Adam and Methuselah, with whom the art of magic began," he wrote, so that the new magic of alchemy ("the way of preserving long life") can restore humans to where once Adam and Eve were in the Garden.[42]

WONDER DRUGS

During the thirteenth and early fourteenth centuries, the notion of a possible elixir fired people's imaginations; from this line of thought developed the idea of wonder drugs, just as Alderotti proclaimed. In the thirteenth century and for centuries thereafter, Galenic pharmaceutical theory gradually was employed both to explain drug actions and to determine theoretically drug administration. Galen proposed a medical concept that was compatible with the Ptolemaic-Aristotelian-Empedoclean cosmic notions about matter and its behavior. Any substance is potentially a drug because of its composition unless it was one of few substances that were inert or neutral because of a perfect balance of qualities. Substances had active (either warming or cooling) and passive (either moistening or drying) qualities. A physician's task is to restore balance to the four humors constituting the body's chemistry. Why is it that a minuscule amount of a drug has little effect, and when one increases the amount and frequency, then the desired effects increase? The increase, however, is only to a point. At that point, increased amounts and their frequencies have deleterious effects. We use the term dosage to explain the observation. According to Galen (who probably was relating an explanation already current among practitioners in his time), each substance has the potential for one of four degrees of activity, for moisture and temperature control. Thus, for example, a fever needs a cooling medicine. An infant might have a higher fever but should be given a lesser amount, whereas a fever in an old person is more threatening and requires a stronger response. Responses are not singular, because the fever-reducing drug would also have one of four degrees of activity to be either moistening or drying. If a fever were wet (sweats), one would need a counterdrying drug. If the drug selected were also moistening, a counterdrug would have to be administered to balance the hydration factor.[43] The complexity in prescribing drugs had the cumulative effect of producing compound drugs with elaborate, mathematically balanced qualities. The growth of compound drugs had an appealing factor that made these mixtures economically gainful to apothecaries. Medicine was less the kitchen herb garden but now bottles sold in apothecary shops in the precursor to our patent drugs.

The wonder-drug concept developed as a response to the complexity of Galenic-drug theory and the idea that there were special substances that could act in a wondrous way to cure virtually every disease and even to control emotions and protect one from harm. The nearest synonym in modern parlance is "silver bullet" (or, variant, "miracle drug"), that is a drug that pathologically cures an intractable disease and, hopefully, pleasantly as well. Francis B. Brévart of the University of Pennsylvania recently studied the "wonder drugs" primarily in German medical and pharmaceutical treatises. Brévart observed that these wonder drugs also possessed occult forces.[44] Wonder drugs combined therapeutic and occult virtues in much the same way as the elixir concept. The difference was one of degree between the two (wonder drugs and elixirs): a wonder drug had specific targeted effects on disease, maladies, and demonic spirits; an elixir extended life in a mysterious way with or without specificity of actions. Even so, Brévart observed: There was "a firm belief in supernatural powers now rendered these objects capable of fulfilling a multiplicity of functions."[45] The concept of wonder drugs was not confined to the elite of natural philosophers who speculated on the nature of things. Brévart's study looked at the "house books" that were popular guides to daily living. Why should they not be appealing? One with a debilitating, life-threatening disease where the local pharmacy was of no use would grasp at the possibility that there was a drug that potentially cured what ailed him. Do we not do this even today?

Artemisia as Example of Wonder Drug

The plant called artemisia or mugwort in our common parlance is a wonder-drug example, all the more so because of its connection with Diana or Artemis. The immensely popular German work *Gart der Gesundheit*/ "Garden of Health" (Mainz 1485) by Johann Wonnecke von Kaub related mugwort's pharmaceutical, magical, and apotropaic qualities, all mixed together in a wondrous way. Wonnecke gave as its synonym "*Mater herbarum*/ Mother herb" and traced its name to King Mausolus' wife, but he also included Isidore of Seville's assertion that the plant was consecrated to Diana "by the pagans."[46] Francis B. Brévart observed: "Wonnecke's introduction of Artemis (Diana), the goddess of childbirth and fertility, the protectress of women, at the onset of the treatise would serve to instill all the more confidence in the virtues of this magiferous herb about to be enumerated."[47] Wonnecke related well the medicinal values for the plant as he received them from his classical and medical sources, but there is a greater emphasis on the wondrous or magical qualities. For example, Wonnecke wrote: "Dioscorides, once again, in his chapter on artemisia informs us that an individual using mugwort cannot be harmed by poisons or sorcery, nor can be injured by the bite of a sick animal."[48] He stretched Dioscorides, who said that, drunk with wine, mugwort is an antidote for deadly poisons and an application with olive oil for "shiverers." It could be used in a litter or as a fumigant to chase away serpents and, when drunk with wine, it protects against poisonous spiders and scorpions.[49] Wonnecke's rendering of Dioscorides conveys the impression that mere possession of the plant protected one from poisons.[50] This inference is made stronger by the two passages that say: "[9] Also, whoever keeps mugwort inside his house will not suffer any damage from the devil. [10] And whoever suspends mugwort root from the neck will not suffer any injury from a venomous animal. And if such an animal has already attacked him, let him drink mugwort juice, and he will regain his health momentarily."[51] Also claiming Dioscorides as his source, Wonnecke said that one who carries mugwort will not have fatigue or, if one hangs it over his house's threshold, he will be protected from harm.[52] Dioscorides's Greek text has no such claims, but a number of pseudo-Dioscoridean texts circulated in the Middle Ages. To be noted is that Wonnecke neither made an explicit tie of mugwort to an astrological sign that would be expected in a Hermetic-inspired work nor did he have favorable calendar times or a ritual for harvesting the plant. One of Wonnecke's sources, the Middle High German *Macer floridus,* gave instructions that leaves should be cut downward, in what we suppose was a magic belief that the downward cut would produce the downward effect for menstrual problems, thus its "signature."[53] Female cures and treatments account for the high plurality of pharmaceutical benefits. The numerous listings of artemisia's medicinal qualities, most with a basis in fact as we know them, were combined with these mystical (magic and apotropaic) qualities to produce a wonder drug, one that cut through the restrictions of nature imposed by the limitations of mere matter composed of water, earth, fire, and air. Herbs once connected with the gods broke the bonds of nature's laws.

WITCHES, WITCHCRAFT, AND DRUGS

From the fourteenth through the seventeenth century in Europe and America approximately one-half million people, more than 80 percent women, were burned at the stake as witches. Some few of the accused were exonerated, after confession under torture. Most were not. Wrestling with a complexity of questions, historians pose these questions: were these witches "really witches," or was the phenomenon of witchcraft suppression a cruel hoax visited upon innocent people whose only weakness was confession to witchcraft practices because of the dreaded agony of torture? The weight of the evidence supports an answer: at least some people believed themselves to be witches. This thesis moves us to the next question: what do we mean by a witch? Lilith, the old woman who made her home in the bough of the *huluppu* tree, was considered to have been the first witch. Later, she was depicted in Sumerian art as a barren old woman with wings and taloned hands and feet; standing beside her were owls and lions (see figure 6.1). The owl symbol refers to her snatching babies at night. Also she seduced men while they slept and drank thereafter of their blood. To the Jews, she was (see chapter two) Adam's first wife. Isaiah 34:14 describes the day of vengeance: "The wild-cat shall meet with the jackals. And the satyr shall cry to his fellow; yea, Lilith [*Liyliyth*] shall repose there."[54] The Greek Septuagint rendered *Liyliyth* into *onokentamos,* meaning "demon," whereas the Greek Aquila translation was *Lilith.*[55] In Greco-Roman times Lilith was Lamia who lost her children and thereafter tore babies from their mothers' arms.[56] Medieval people's emphasis on Lilith was as a snatcher of babies before baptism, thereby removing neonatals' possibility of salvation. A German woodcut in 1470 shows Lilith with wings and a serpent tail offering an apple to Eve. This image connects evil with Lilith (partial serpent, baby-snatching) evil, and the fall from Grace, all in one simple scene (see figure 6.2).

If witches were Lilith-like spirits, they were assumed to exist. If medieval witches were nothing more than sorcerers, they had existed from antiquity and without interruption through the Western medieval periods. If, on the other hand, witches are defined as engaging in formalized, ritualistic worship of the Devil, the historical controversy becomes muddled and returns to the question: Were all those witchcraft confessions real or perpetrated upon innocent victims through torture? Partially these associations were an outgrowth of the early medieval belief in demons. One tradition regarded demons as fallen angels; the devil fell from grace because he

Figure 6.1 Lilith with owls and lions on each side on a Babylonian cult plaque, ca. 1950 BCE.
Source: The Burney Relief, The Granger Collection, New York (0017755).

Figure 6.2 Lilith tempting Eve with an apple in the Garden of Eden.
Source: Woodcut, German, 1470. The Granger Collection, New York (0044522).

envied Adam and his temptation of Eve. Another tradition held that God commanded angels to worship Adam and, when they refused, they were cast down from heaven. [57] As bodiless spirits they could occupy and possess a human. There were many variations and manifestations of these stories, but the salient point is that from peasants to high churchmen in both Western and Eastern Christendom, many people believed in the existence of demons in their everyday lives. Witchcraft came from the devil's placing in humans anger and a "desire to do harm," as stated by Roger Bacon.[58]

The main parameters of a witch-convent ritual, according to confessions went like this: On a Thursday (or Tuesday, with regional differences) a person's spirit (usually a woman) arose from her bed, rubbed herself with green unguent (often described as foul-smelling) and flew through the window, wall or roof and was transported (often on a stick or broom) to the woods, where others were arriving in similar ways. Naked, they formed a circle linking their arms and facing outward. They danced, and at the height of frenzied dance the devil appeared in the center, where they gave homage (the devil's kiss, as often stated whose details I shall spare). Then they had a banquet, gave pledges to act in accordance with the devil's precepts, engaged in an orgy, and returned home just as they had come. Since it was their spirits, not their bodies, that had been thus transported, the testimony of a spouse was immaterial inasmuch as the body remained in bed. In the fifteenth century, a major accusation against witches was that they killed babies and small children; this became a common theme. Supposedly some witches could steal into a home at night while parents were asleep and choke the children to death.[59] Historians cannot discern from the confessions what true sentiments of the villagers and towns people who allegedly made these claims were.

It was not until the sixteenth century that a number of writers, among them herbalists, gave the formula for the green unguent: it consisted of a variety of herbs, those locally available. All of the recipes contained plants, mandrake and/or hemlock, nightshade, and jimson weed, all with similar chemistries. A sixteenth-century physician in Spain found the formula and tested it on a woman who became comatose for more than a day, around thirty-six hours. When she was revived finally after a great scare that she would not recover, she was upset because she was awakened from pleasant memories of handsome young men in sexual fantasies. Another physician in France described those women who smeared the ointment over themselves and remembered later "dancing, handsome young men, and lovemaking of whatever kind they most desire."[60] Atropine, the active compound in mandrake and some other related Solanaceae plants, has a similar effect when administered topically, of a comatose state for a number of hours, some exceeding a twenty-four. The compound in an ointment is most highly absorptive through the vagina and, when administered on a stick (as one might well do), one of the first visions reported by those so treated is flying through the air on sticks.[61] Like many hallucinatory drugs, there is an out-of-the-body experience. Johann Weyer (1533) tested the green unguent on himself and argued that witches should be excused for their activities because their fantasies were all attributed to drugs and were not real. His conclusions about witchcraft were

> They frequently abuse of the natural remedies which are given to them for their comfort...They think they see theatres, beautiful gardens, feasts, beautiful ornaments, clothes, handsome young men, kings, magistrates, in short, all those things which delight them and which they thus believe themselves to be enjoying; they also see devils, ravens, prisons, deserts, and other torments. They are therefore the causes of violent dreams.[62]

Note that Weyer wrote from the assumption that those who partook of these hallucinatory drugs were women.

Maxim Weintraub compared witchcraft accounts regarding the green unguent with modern clinical accounts of atropine toxicity (included subcompounds of hyoscine and hyoscyamine). He concluded that

> Implicit within most of these tales is one or more of the following aspects: that the person was asleep, experienced sensations of flight, and often believed that the experience actually occurred. These symptoms are consistent with the twentieth-century clinical description of hyoscine and hyoscyamine toxicity; specifically coma, visual, auditory, and tactile hallucinations and the confusion between reality and drug-induced perceptual distortions.[63]

An historian is not justified in concluding that witchcraft was merely a drug culture whose participants' visions the Inquisitors and other witch suppressors misunderstood. There were too many witnesses to strange rituals in the woods and accounts of witnesses of rituals to counter a theory that witchcraft/devil worship was entirely a matter of minds under drug control. Moreover, one who has mandrake and other related drugs would not have motor skills to enable dance and movements reported by those who saw the rituals.[64] One could argue, not persuasively to me from the evidence, that the witnesses were not reliable. There were too many over too long a period with too many variations to be entirely inventive. What transpired at the village level, however, was one thing to the villagers, another to the suppressors. We know what the suppressors thought, but finding the voice of the villagers is more difficult.

Instruction about witchcraft comes from a work (first published in 1467) by two Dominican monks whose purpose was to instruct magistrates how to identify, question, and convict witches. "What do witches do?" was a question that these monks posed, citing papal authority. First, however, the heading was "What sort of women are found to be above all others superstitious and witches?" Witches' acts were sevenfold:

1. Inclining the minds of men to inordinate passion
2. Obstructing the generative force
3. Removing "the member" [i.e., the penis] accommodated to that act
4. Changing men into beasts by their magic art
5. Destroying the generative force in women
6. Procuring abortions
7. Offering children to the devils.[65]

Of the seven, six are directly linked to reproduction and sex. The chaste tree was, to our knowledge, the only substance that was available, which could produce some of these seven effects. Allegedly, witches tied an invisible ligature (string) around a man's penis to prevent an erection. Only the chaste tree could produce effects one through three. The last three, five through seven, concern birth control and snatching neonatals before they can be baptized, the alleged act of Lilith. Abortion was explicit, contraception implicit. The monks cite St. Augustine's *City of God* (bk. 36) as saying: "Devils are attracted by various kinds of stones, herbs, trees, animals, songs, and instruments of music... [and by] spirits by signs."[66] These phenomena (accusation, in this case) are found in Hermetic and alchemical works, albeit, not for demons' evil purposes. The existence of devils (or demons) pervades the Hermetic works. Witches were accused of enacting "black magic," that is for evil purposes, not benign ones. Those who argue that men "can work witchcraft through certain herbs and occult causes without the help of devils" are wrong because devils, not men, produce "corporeal effects." Men who have a "frantic infatuation of love" are beset by a devilish spell. "The natural reason is that she [a woman] is more carnal than a man": so stated these two fifteenth-century misogynist monks.[67]

The monks gave little detail on what the herbs and stones were but we can easily extrapolate that some were the sex-related herbs we have discussed: mandrake, artemisia, and chaste tree.[68] Specifically the authors pointed to "absinthium" (wormwood or artemisia) because of its biblical (Proverbs 5:3–4) reference to a "loose woman" with speech as "bitter as wormwood." In the medieval and early-modern periods, the contraceptive use of pomegranates was only occasional throughout the documents. I speculate that the reason was breeding for tastes had to some degree altered the pomegranate chemistry, thereby reducing the most potent estrogenic compounds. There were more effective contraceptive and abortifacient plants, as I recently offered as evidence in another study.[69]

Who knew these herbs of devilish use? It was the "midwives, who surpass all others in wickedness."[70] Generally, also, women cared for the herb gardens, which provided most medical care in families. Throughout antiquity, the Middle Ages, and the Renaissance, women typically received their extra-family health care from midwives as general advisers on medical care. Henry Boguet's *Examen of Witches* (in the 1580s) spoke of

> those midwives and wise women who are witches are in the habit of offering to Satan the little children which they deliver, and then of killing them... [or] worse; they kill them while they are yet in their mother's womb. This practice is common to all witches.[71]

Jean Spot, a sixteenth-century alleged witch, knew the "empirical remedies/ *remedia empirica*" but she also knew "magical remedies/ *remedia magica*." She practiced a ritual to promote fertility when done in the sun, but, when she went "against the sun," she had evil intent for her remedies. A women testifying before the Inquisition in Modena (1499) said it succinctly: "Who knows how to heal knows how to destroy."[72] However we view the witchcraft suppressions and attendant hysteria, this much is clear: the persecutions of midwives were not delusions of drug-crazed minds either of those of the midwives or of their persecutors. The persecutions were real, and they served useful purposes, from the perspective of those who persecuted them. Most who died because of accusations of witchcraft were doubtlessly innocents, but their knowledge of herbs caused their deaths in many, if not most, cases.

Why Witchcraft and Herbal Lore Suppression

Two German scholars, Gunnar Heinsohn and Otto Steiger, compiled data that showed that in those areas of Europe where there were virulent witch hunts, there followed a period of population increase.[73] Much of late medieval government was through town or municipal government. Guild leaders—for it was they who largely controlled public life—were acutely aware that economic prosperity was linked to growth in town size. Fewer people meant lower rates of income, and the converse. Who were the suspects for lower population growth?: women acting on the advice of midwives. For that reason some towns outlawed the growing and possession of juniper, a widely used early abortifacient, in what was probably the first law by a governmental entity to regulate substance use and possession. The laws did not work: no matter how dedicated the police were, people were industrious enough to hide their juniper bushes. One method worked, if the Heinsohn-Steiger thesis is correct: the extermination of midwives and those who dispensed herbs and advice on how to use them to curtail fertility and live births resulted in fewer people having the knowledge.

Another means of testing the thesis about the link between population and witchcraft suppression comes with the coincidence of chronology. Some historians (e.g., Margaret Murray and Arne Runeberg[74]) believe medieval witchcraft was a direct descendant of the ancient fertility cults, modified by Christian thought but not substantially changed from its ancient roots. Without doubt, throughout the Middle Ages, even as a continuation from classical antiquity, there were or were thought to be sorcerers and magicians. It was never that people and the Christian churchmen and women did not believe in magic: they accepted on faith the miracles of saints and of divine intervention. Magic for evil purposes was their dread. For example, Gratian, a well-known canon lawyer, wrote around 1140: "Bishops and their officials must labor with all their strength to uproot thoroughly from their parishes the pernicious art of sorcery and malefice invented by the Devil, and, if they find a man or woman follower of this wickedness to eject them foully disgraced from their parishes."[75] The key question for the historian is why, when there were magicians and sorcerers, even those allegedly serving the devil, living throughout the medieval period, did Western Europe change and no longer to tolerate them? In 1320 Pope John XXII instructed a cardinal to have the Inquisition prosecute sorcerers as heretics. Some historians argue that the Inquisition was an institution; others say the term is too precise for a loose organization evolving from the control of bishopric or other ecclesiastical authorities into virtual operational independence. Not until about the 1420s did the Inquisition evolve out of any central control or authority into the terrible oppression of witchcraft.

"Thou shalt not suffer a witch to live," so read the King James translation of Exodus 22:18. The Latin and Greek bibles used a term meaning "sorceress" or "poisoner"; the essential point is that this passage was known as God's word throughout the Middle Ages, yet it was not until the end of the medieval period that the words evoked actions. Why? History provides an answer: In the twelfth century conventional Western wisdom held that right reason will bring all to accept God and Jesus as his Son. After more than a hundred fruitless years of crusader armies marching off to the Holy Land, the few returning reported that the Saracens held to the Prophet, not Jesus, and would not convert by persuasion. Europe faltered in asserting that reason would bring all people to the same conclusions. Also, the Westerners saw the Muslims as a threat largely because of their numbers. They knew two things: Muslim men could have multiple wives and many children. After the depression beginning in the 1320s and cataclysmic debacle of the Black Death (beginning in 1347 and recurring in episodes thereafter), some Europeans and even theologians argued that Christians should drop their monogamy for self-preservation. Whereas the early Christian Church emphasized virginity and chastity, the late medieval Church entertained policies to enhance population size because of insufficiencies of multiplication.[76]

Witches were seen as those who prevented babies from being born through contraception, abortion, snatching of non-baptized neonatal babies, causing men to be unable to have an erection (impotence), and variously "obstructing the generative act": so spoke two of its accusers quoted earlier. Hincmar of Rheims (d. 882) discussed *striga* (witch) who used magic (*maleficium*) to destroy marital affection.[77] In classical Latin a *strix* meant an owl, later meaning also a night-spirit and vampire.

How did the witches allegedly enact their mischief: through herbs, potions (the usual term), spells, and enchantments. Spells and enchantments could be subsumed under empowering drug actions as well as such bizarre accusations as causing hail-storms or floods. Who knew how these drugs worked, how to find, gather, and prepare them, how much to give and how frequently? The answer to these questions is midwives and old women. To a degree the accusations were true. Heinrich Marzell, a German scholar, demonstrated that certain herbs were associated with witchcraft. He named them "Zauberpflanzen Hexentränke/ Magic plants and witch potions" and "Hexen- und Teufelskraüter/ Witches' and Devils' Herbs."[78] He included specifically mandrake and artemisia and other mind-affecting herbal plants such as St. John's wort. Any woman who possessed these devilish herbs or even had knowledge of them would be suspected of being a witch. "The better the midwife the better the witch," so was a statement traced to the Middle Ages.[79] If one possessed one or more of these herbs or even had the reputation of knowing them, she would be a likely witch suspect and, in the frenzy of a virulent witch-hunt by zealous, fanatical inquisitors, the possession or knowledge would be sufficient for a trial. The knowledge of these birth-controlling and mind-altering drugs was virtually erased from folk memory, and such knowledge was not part of formal medical training in the universities. A physician was taught how to heal, not to poison.

Words used to designate witches in the Middle Ages indicate just how close the association was between sorcery, witchcraft, poisoning, midwives, and herbalists. The words for "wise woman" was virtually synonymous with midwives: *vetula* ("old woman"), *mulier* ("woman"), *obstetrix* ("obstetrician"), *Weise Frau* ("wise woman"), *saga* ("wise woman"), *saga matrona* ("wise woman"), and *sage-femme* ("wise woman"). These words indicated a possible witch, that is to say, in fourteenth-century parlance, suspected of being a devil worshipper or sorceress. Two Germanic words, *Nacht Frau* ("night woman") and *Feld Frau* ("Country Woman") have a similar connotation. The connection between witches and poisoning is seen in these words also indicating witch: *venefica* ("poison woman"), and *Gifft-Köche* (poison-cooker, *f.*). The word for female herbalist, *herbaria*, indicated a witch. The English word "witch" in Middle English was "*wicca*" (masculine) and used as early as about 890 in the Laws of Alfred,[80] which threatened witches with death on the basis of the Exodus passage quoted earlier. Edward the Elder (d. 924) exiled all who were *wiccan* and *wigleras* ("soothsayers"), and his successor, Ethelstan, executed all *wiccecraeftum* if their actions resulted in death. As early as 1000 the feminine form, *wicce*, was frequently used for the designation.[81]

Pertinent to our hypothesis about gods, herbs, and witches is the number of words for witches that point to a bird: *Strix*, as we saw above in the Hincmar quotation, was in classical Latin an owl or screech owl. Later medieval forms were *strega, striga,* or *strigimaga* that meant witch, usually female, with a connotation of a screech owl.[82] A modern source says that *strega* was a vampire that could turn into an owl and later was applied only to female witches.[83] The changing into beasts probably was the last act of the sevenfold witchcraft, cited above (No. 4: "Changing men into beasts by their magic art") for which herbs were not directly responsible. Similarly, *scobax* refers to both an owl and a witch as a transfer from the Greek, *skopō.* In classical Greek this kind of owl that was more nocturnal than the Athenian owl was called in scientific nomenclature *Athene noctua.*[84] Interestingly the science name for *skopō* is *Strix scops L.* or "the witch owl." Linnaeus gave the owl its name in Latin apparently knowing its connection with witchcraft. Today the European horn-owl is known as scops. One of the earliest uses of the term for witchcraft (*wiccecraefte*) in English occurs is a work called *Owl and Night* about 1250.[85] In chapter one, we speculated that the *Anzû* might have been an owl. Earlier in this chapter, depictions of Inanna show her as an owl that snatched babies at night. In the same chapter, we learned that Ascelaphos, the god who turned in Persephone for eating the pomegranate, was turned into a screech owl.[86]

Conclusion

From the earliest time when humans in a complex urban socially stratified society left artifacts and writings, goddesses and herbs were strongly connected. Some of the goddesses' powers were thought to be enacted through specific herbs. Even biblical works, while denying the goddesses, attested to their uses. The pomegranate, mandrake, artemisia, and chaste plant are prominent examples, but there were other herbs. Only through Chinese science, which respects traditional folk remedies, did we return to artemisia plants for curing malaria. The medicinal and hallucinatory effects attributed to these herbs were real, as judged by our science. Why did we lose much of this knowledge? After the Western world became Christianized, four movements blocked the link between herbal medicine and human usage: the Hermetic philosophy that practical knowledge was useful according to a divine plan; the spread of the imaginative Chinese concept of an elixir to confer prolonged life; the attendant Western development of the existence of wonder drugs; and, finally, the demonization of these miraculous drug-supplying herbs when mere knowledge of them could be a death sentence. The knowledge given to humans by the gods became the instrument of demons and the devil. Inanna, Artemis, Demeter, and Hermes would have been appalled to learn what happened to their herbs. Adam, Eve, John Donne, Robert Service, Vincent Van Gogh, and hundreds of thousands so-called witches knew all too well that god's herbs could have evil consequences. Also the herbs could work wondrous works to behold. Even now, we cherish, hope and believe that there will be a wonder (*variant,* "silver bullet" or "miracle") drug to cure cancer, Alzheimer, and even cardiac diseases and that some day we will find the elixir. As I write this, attention for the elixir turns toward a compound in red wine—not a bad place to look—but, by the time you read this, probably the search will be looking elsewhere. It is out there; we need to only find it, just as Hermes told us. What a shame that in our search, we lost so many of the clues that the past provides for us.

NOTES

INTRODUCTION

1. John Riddle, "Folk Tradition and Folk Medicine: Recognition of Drugs in Classical Antiquity," in *Folklore and Folk Medicines*. John Scarborough, ed. (Madison: American Institute of the History of Pharmacy, 1987), 33–61.

CHAPTER 1

1. As quoted by David Aaronovitch in *The Guardian*, Tuesday June 10, 2003 http://education.guardian.co.uk/higher/comment/story/0,,974440,00.html (accessed 4/16/08).
2. Ibid.
3. Matthew Bogdanos with William Patrick, *Thieves of Baghdad* (New York: Bloomsbury, 2005), esp. p. 136, where Bogdanos notes the poor inventory on cards of what was there and how that interfered with a precise records of recovery.
4. Selma Al-Radi, "The Ravages of War and the Challenge of Reconstruction," in *The Looting of the Iraq Museum, Baghdad, The Lost Legacy of Ancient Mesopotamia*. Milbry Polk and Angela M. H. Schuster, eds. (New York: Harry N. Abrams, 2005), p. 210 (hereinafter cited as *Lost Legacy*); Bogdanos, p. 270.
5. Lamia Al-Gailani Werr, "A Museum Is Born," in *Lost Legacy*, pp. 27–28.
6. Denise Schmandt-Besserat, *When Writing Met Art: From Symbol to Story* (Austin: University of Texas Press, 2007), p. 36; see also description by Zainab Bahrani, *Women of Babylon: Gender and Representation in Mesopotamia* (London: Routledge, 2001), pp. 134–140 and Charles S. Maier, *Among Empires: American Ascendancy and Its Predecessors* (Cambridge, MA: Harvard University Press, 2006), pp. 285–295 (on Uruk vase).
7. C. Wicke, "Inanna/Ištar," in *Reallexikon der Assyriologie und vorderasiatischen Archäologie* (Berlin: Walter de Gruyter, 1980), 5: 74–87; and iconographic aspects by U. Seidl, 5: 87–89.
8. I.S.S. [*sic*] *Nuptial Rites, Or the Several Marriage Ceremonies Practiced amongst all the Nations in the World* (London: "Printed by T. S. for the Author," 1685), p. 27; BBC h2g2, "Popular Wedding Traditions and Superstitions," online http://www.bbc.co.uk/dna/h2g2/A3383633 (accessed 11/2/08).
9. For examples: http://www.farsinet.com/persian_wedding (accessed 2/26/08); http://www.globosapiens.net/travel-information/Esfahan-1974.html (accessed 2/26/08); and http://www.billcasselman.com/wording_room/pomegranate.htm (accessed 2/26/08) for Greek weddings.

10. Schmandt-Besserat, p. 1 (and specifically on the Uruk vase, 41–46). One wonders whether there is possible connection between the myth and custom in the Attic marriage ceremony. Following the wedding, the bride went to the groom's house and ate a quince upon arriving. Was there a substitution of quince for pomegranate? If so, what meaning may the innovative substitute have had? Was there a new message delivered by Attic brides to Attic husbands?
11. Mary K. Wakeman, "Ancient Sumer and the Women's Movement," *Journal of Feminist Studies in Religion* 1/2 (1985): 7–27.
12. Thorkild Jacobsen, *The Treasures of Darkness: A History of Mesopotamian Religion* (New Haven, CT: Yale University Press, 1976), pp. 16, 25–47.
13. J. Van Dijk, "Le motif comique dans la pensée sumérienne," *Acta Orientalis* 28/2 (1964): 1–59.
14. Schmandt-Besserat, pp. 4–12.
15. Wakeman, 12–13, 20–23; Elizabeth Gould Davis, *The First Sex* (New York: G. P. Putnam, 1971), pp. 44–72.
16. "Sumerian History, Culture and Literature," in *Inanna Queen of Heaven and Earth: Her Stories and Hymns from Sumer.* Diane Wolkstein and Samuel Noah Kramer, eds. (New York: Harper & Row, 1983), pp. 124–125 [115–135].
17. Samuel Noah Kramer, *The Sacred Marriage Rite: Aspects of Faith, Myth, and Ritual in Ancient Sumer* (Bloomington: Indiana University Press, 1969), p. 57.
18. Samuel Noah Kramer, "The Discovery and Decipherment of the Descent of Inanna," in *Inanna*, pp. 127–135.
19. Nicola Vulpe, "Irony and the Unity of the *Gilgamesh Epic*," *Journal of Near Eastern Studies* 53/4 (1994): 275–283.
20. From "Descent of Inanna," in *Inanna,* p. 52; a more recent publication of the translated legends appears in *Inanna from the Myths of Ancient Sumer* by Kim Echlin and illustrated by Linda Wolfsgruber (Toronto: Groundwood Books, 2003), which culls from older translations. The author explains that she condensed some material and published the work as a reader for her daughters. The illustrations, while modernized, were well drawn and attractive.
21. Wolkstein, "Interpretations of Inanna's Stories and Hymns," in *Inanna*, p. 156.
22. Vulpe, pp. 279–280.
23. Wolkstein in *Inanna*, p. xvi.
24. "Descent of Inanna," in *Inanna*, p. 55.
25. On Inanna' early relationship with Dumuzi, see Thorkild Jacobsen, *Toward the Image of Tammuz and Other Essays on Mesopotamian History and Culture.* William L. Moran, ed. (Cambridge, MA: Harvard University Press, 1970), pp. 52–72, 83–101.
26. Ibid., p. 60.
27. Jacobsen, *Treasures of Darkness,* p. 57.
28. Wolkstein, "Inanna and the God of Wisdom," in *Inanna*, p. 12. Wolkstein (p. 146) explains that the Sumerian word that she translates as "vulva" also can mean "sheepfold," "womb," and "loins."
29. Richard A. Henshaw, *Female and Male: The Cultic Personnel: The Bible and the Rest of the Ancient near East* (Allison Park, PA: Pickwick, 1994), p. 5.
30. Ibid.; Wolkstein and Kramer, p. 56, with translation: "Come, man, come!"
31. Wolkstein, "Interpretations," p. 139.
32. Matrifocus in http://www.matrifocus.com/BEL05/spotlight.htm (accessed 3/17/08).
33. Translation by Samuel Noah Kramer in *Inanna*, p. 5.
34. Ibid.
35. Wolkstein, "Interpretations," p. 142.

36. *Inanna,* p. 6.
37. Ibid., p. 9
38. Jeffrey H. Tigay, *The Evolution of the Gilgamesh Epic* (Philadelphia: University of Pennsylvania Press, 1982), p. 190.
39. A. R. George, *The Babylonian Gilgamesh Epic,* 2 vols. (Oxford: Oxford University Press, 2002), p. 898.
40. A. Leo Oppenheim.*The Assyrian Dictionary of the Oriental Institute of the University of Chicago* (Chicago: Oriental Institute and Gluckstadt: J. J. Augustin), 9 (1956): 190. (hereinafter cited as *CAD)*; Franz Köcher, *Die babylonisch-assyrische Medizin in Texten und Untersuchungen,* 6 vols. (Berlin: Walter de Gruyter, 1963–1980). (hereinafter cited as *MTU*): *Reallexikon der Assyriologie und vorderasiatischen Archäologie* (Berlin: Walter de Gruyter, 1980), 4 (1992): 324–325 (hereinafter cited as *ABD*); Richard P. H. Greenfield, *Traditions of Belief in Late Byzantine Demonology* (Amsterdam: Hakkert, 1988), p. 185, identifying the Assyrian demon as "Lamashu."
41. Talmud: *Nidah.* 24b (in *Seder Tohoroth,* Epstein trans. 1:66); *Baba Batra.* 73a (*Seder Nazikim,* 2:290); *Sabbath.* 151b (*Seder Mo`ed,* 1: 773; *'Erubin,* 100b. (*Seder Mo`ed.* 2:698); Lowell K. Handy, "Lilith," *Anchor Bible Dictionary,* 4 (1992): 324–325.
42. Louis Ginzberg, *The Legends of the Jews.* Henrietta Szold, trans. from German. 4 vols. (Philadelphia, PA: Jewish Publication Society of America, 1909) 1: 65–66; Selma R. Williams, *Riding the Nightmare: Women and Witchcraft from the Old World to Colonial Salem* (New York: Harper Perennial, 1978), p. 76; Angelo S. Rappoport, *Ancient Israel: Myths and Legends* (New York: Bonanza Books, 1987), pp. 77–79.
43. Handy in *ABD*, p. 325.
44. Sergio Ribichini, "Lilith nell-albero H-uluppu," *Atti del l'Convegno Italiano sul vicino oriente antico (Roma, 22–24 Aprile 1976).* Orientis Antique Collection 13, Rome, 1978, pp. 25–33.
45. *CAD* 1: 153–155.
46. Electronic Text Corpus of Sumerian Literature: http://www.orinst.ox.ac.uk/cgi-bin/etcsl (accessed 3/2/08).
47. Amar Annus, *The Standard Babylonian Epic of Anzu* (State Archives of Assyrian Cuneiform Texts, vol. 3) (Helsinki: Neo-Assyrian Text Corpus Project, 2001), p. xxv, and Amar Annus's Web site with an interpretive essay in http://www.gatewaystobabylon.com/essays/wisdomninurta.html (accessed 3/20/08).
48. Electronic Text of Corpus of Sumerian Literature. Ninurta Epic, c. 1. 6. 1 in http://etcsl.orinst.ox.ac.uk/cgi-bin/etcsl.cgi?simplesearchword=Anzu&simplesearch=translat... (accessed 3/2/08). For the cuneiform text and commentary see, Annus.
49. Ovid, *Fasti,* James G. Frazer, trans. (Cambridge, MA: Harvard University Press; London: Heinemann, 1951), 5: 533–550.
50. See chapter six of this book for the owl association with witchcraft.
51. Another Sumerian story is of Lugalbanda, the hero, and the Thunderbird, who raises its young in the mountains, preys on oxen with its large size, and possesses magical powers. In the story, the thunderbird seems to have lost relations with the rain and thunder and, according to Thorkild Jacobsen (*The Harps that Once... Sumerian Poetry in Translation* [New Haven, CT: Yale University Press, 1987], p. 321) is a "mere fairy-tale." Greenfield (p. 185) relates that the *anzû* was associated with Tiamat in the *Enuma elish* creation story and the Hebrew Tehom, the evil mother of demons.
52. Henshaw, pp. 173–174.
53. Acts 28: 3–4 (Revised Standard Translation).

54. Henshaw, p. 6.
55. Wolkstein and Kramer, p. 178, but elsewhere Kramer (*The Sumerians: Their History, Culture, and Character* [Chicago: University of Chicago Press, 1963], p. 198) believes it may be a willow and, without explanation, A. R. George (*The Babylonian Gilgamesh Epic*, 2 vols. [Oxford: Oxford University Press, 2002] 1:3) identifies the tree as a willow.
56. Marvin A. Powell, "Timber Production in Presargonic Lagaš," in *Trees and Timber in Mesopotamia* ("Bulletin on Sumerian Agriculture," vol. 6, Cambridge, 1992), p. 109.
57. J. N. Postgate, "Trees and Timber in the Assyrian Texts," in *Trees and Timber*, pp. 177, 179, and, especially, 182.
58. *CAD* 6 (1956): 55–56. (hereinafter cited as *CAD*); Postgate, "Trees and Timber," p. 182, says that he knows no basis on which the claim made in the *CAD* could have been made pointing to the oak.
59. G. Wilcox, "Timber and Trees: Ancient Exploitation in the Middle East: Evidence from Plant Remains," in *Trees and Timber*, pp. 6, 17.
60. Marc Van de Mieroop, "Wood in the Old Babylonian Texts from Southern Babylonia," in *Trees and Timber*, p. 159, who doubts that oak is the correct identification; M. A. Powell, "The Tree Section of *ur5(=HAR)-ra=hubullu*," *Bulletin on Sumerian Agriculture* 3 (1987): 146, 148 [145–151], who puts *haluppu* as an unidentified "exotic tree," possibly an oak.
61. *CAD* 6: 56; the manufacture of wood boats in modern southern Iraq is examined by Edward Ochsenschlager, "Ethnographic Evidence for Wood, Boats, Bitumen and Reeds in Southern Iraq," in *Trees and Timber*, pp. 47–78. Mulberry and acacia are more commonly used in boat construction.
62. Ibid., 6: 56.
63. Jacobsen, *Treasures of Darkness*, pp. 20, 26–73.
64. Helgard Balz-Cochois, *Inanna. Wesenbild und Kult einer unmütterlichen Göttin* (Gütersloh: Güttersloher Verlagshaus Gerd Mohn, 1992), Studien zum Verstehen fremder Religionen, vol. 4, p. 48.
65. de Mieroop, p. 159.
66. Ibid., p. 156.
67. JoAnn Scurlock and Burton R. Andersen, trans. and commentary, *Diagnoses in Assyrian and Babylonian Medical Analyses* (Urbana: University of Illinois Press, 2005), pp. 559–560.
68. Scurlock and Andersen say that the warning came in context of love magic and which they placed in the category of a desire to live in a dire condition (p. 560).
69. Ibid.
70. Markham J. Geller, *Renal and Rectal Disease Texts.* Vol. 7 of *MTU* (Berlin: de Gruyter, 2005), 7: 44–45, 98–99.
71. Pankaj Oudhia, "Research Note," in http:botanical.com/site/column_poudhis/77_gyn.html (accessed 3/21/08; site Botanical.com is by subscription).
72. See R. Campbell Thompson, *Dictionary of Assyrian Botany* [*DAB*] (London: British Academy, 1949), p. 315, 11, and Ebeling, *KAR*, 194, iv, 18. Thompson notes the uncertainty of identification of the part of the pomegranate applied to the wool, although he refers to its use in treating a "woman's disease." He renders the term, we believe meaning vagina, as uterus. For parallel uses of pomegranate in the treatment of disease, see *CAD*, 11, pt. 2, p. 346b and *KAR* 192, r, 17 (application of the "flour of the fruit," *ZI'D inbi*, to the sick place). It should also be noted that Thompson's identification of the designation *GIŠ.NU.U'R.MA.KU7.KU7* as a "species" of

pomegranate is probably incorrect. It appears to identify a taste associated with the pomegranate. See *CAD* 8, p. 494b.

73. Navindra P. Seeram, Risa N. Schulman, and David Heber, eds., *Pomegranates: Ancient Roots to Modern Medicine* (Boca Raton, FL: CRC Taylor & Francis Press, 2006); see also my (John Riddle) review in *Fruit Gardener* 39/1 (2007): 26, 28.
74. Diane M. Harris, Emily Besselink, and Navindra P. Seeram, "Assessment of Estrogenicity of Pomegrante in an *In Vitro* Bioassay," in *Pomegranates*, p. 144.
75. E. Lansky, S. Shubert, and I. Neeman, "Pharmacological and Therapeutic Properties of Pomegranate," CIHEAM: *Options Méditerranéennes*, pp. 231–235 [see full citation in bibliography].
76. Michelle P. Warren, Eliza Ng, Russalind H. Ramos, and Sari Halpert, "Absence of Significant Estrogenic Effects in the Postmenopausal Population," in *Pomegranates*, pp. 157–164 (quotations from pp. 161 and 164).
77. Erich Heftmann, Shui-Tze Ko, and Raymond D. Bennett, "Identification of Estrone in Pomegranate Seeds," *Phytochemistry* 5 (1966): 1337–1339; P. D. G. Dean, D. Exley, and T. W. Goodwin, "Steroid Oestrogens in Plants: Re-estimations of Oestrone in Pomegranate Seeds," *Phytochemistry* 10 (1971): 2215–2216.
78. M. L. Gujral, D. R. Varma, and K. N. Sareen, "Oral Contraceptives. Part I: Preliminary Observations on the Antifertility Effect of Some Indigenous Drugs," *Indian Journal of Medical Research* 48 (1960): 50 [46–51].
79. Anand O. Prakash, "Potentialities of Some Indigeneous Plants for Antifertility Activity," *International Journal of Crude Drug Research* 24 (1986): 21, 23 [19–24].
80. Anand O. Prakash et al., "Anti-implantation Activity of Some Indigenous Plants in Rats," *Acta Europaea Fertilitas* 16 (1985): 447 [441–448]; a number of compounds found in the pericarps is found in H. Satomi, K. Umemura, A. Ueno, T. Hatano, T. Okuda, and T. Noro, "Carbonic Anhydrase Inhibitors from the Pericarps of *Punica granatum* L.," *Biological and Pharmaceutical Bulletin* 16 (1993): 787–790 [through Medline].
81. J. J. Segura, L. H. Morales-Ramos, J. Verde-Star, and D. Guerra, "Inhibición del orecimiento de Entamoeba histolytica y. E. invandens producida por la raiz del granado (*Punica granatum* L.)," *Archivos Investigación Médica* 21 (1990): 235–239.
82. Scurlock and Andersen, p. 261; M. Stoll, *Birth in Babylonia and the Bible: Its Mediterranean Setting* (Groningen: Styx Publications, 2000), p. 38; the best study on Babylonian (mostly Assyrian) pharmacy is by Dietlinde Goltz, *Studies zur altorientalischen und Grieschischen Heilkunde: Therapie—Arzneibereitung—Rezeptstruktur.* Sudhoffs Archiv, vol. 16 (Wiesbaden: Franz Steiner Verlag, 1974).
83. Jean Bottéro, *Everyday Life in Ancient Mesopotamia*. Antonia Nevill, trans. (Baltimore: Johns Hopkins University Press, 2001), p. 198.
84. Joann Scurlock and Dafydd Stephens, "A Ringing Endorsement for Assyro-Babylonian Medicine: The Diagnosis and Treatment of Tinnitus in the 1st Millenium BCE Mesopotamia," *Audiological Medicine* 6 (2008): 9, 15; the tablets were published by E. Ebeling (see ref. in bibliography).
85. Iraq Museum, Baghdad, and described by Friedrich Muthmann, *Der Granatapfel Symbol des Lebens in der alten Welt* (Bern: Abegg-Stiftung, 1982), p. 13.
86. The preword *gis* means wood or tree.
87. Robert D. Biggs, *ŠA`.ZI.GA. Ancient Mesopotamian Potency Incantations* (=*Texts from Cuneiform Sources*, vol. 2, Locust Valley, NY: J. J. Augustin, 1967), p. 70. (hereinafter cited as *TCS* 2).
88. Ibid., p. 71.
89. We say this mindful of Wilfred G. Lambert's admonition ("Gilgamesh in Literature and Art: The Second and First Millennia," in *Monsters and Demons in the Ancient*

and Medieval Worlds [Mainz: Philipp von Zabern, 1987], pp. 37–52) that the literary descriptions of objects date from one period and the artistic representations from another, thereby questioning the accuracy of identifications through artifact representations.

90. Muthmann, pp. 13–15 (pictures, p. 15); cf. J. Börker-Klähn, "Granatapfel," *Reallexikon der Assyriologie*, 3 (1971): 619 [616–630]. (hereinafter cited as *RA*).
91. Ibid.; piece is in Vorderasiatisches Museum, Berlin.
92. Claude F. Schaeffer, "Les fouilles de Ras-Shamra cinquième campagne (printemps 1933)," *Syria* 15 (1934): 124–126 (with pictures) [105–134]; Muthmann, p. 19; in Baghdad, Iraq Museum.
93. Muthmann, p. 19; in Baghdad, Iraq Museum.
94. Louvre, Paris, Inv. S. 1208 (Bj.2169), described by Muthmann, pp. 35–38.
95. Shalmeneser's roll is now in the British Museum, and the other is in the Staatliche Museen in Berlin; reproduced by Muthmann, p. 21.
96. Alabaster relief in Louvre, Paris; Muthmann, pp. 24–25; for other representations of the pomegranate in Assyrian artifacts, see Erika Bleibtreu, *Die Flora der neuassyrischen Reliefs* (Vienna: Institutes für Orientalistik, 1980), pp. 183–186, 213–215.
97. Alasdair Livingstone, *Mystical and Mythological Works of Assyrian and Babylonian Scholars* (Oxford: Clarendon Press, 1986), p. 97.
98. Paul Jacobsthal, *Greek Pins and Their Connections with Europe and Asia* (Oxford: Clarendon Press, 1956), pp. 185–199, esp. 187.
99. Muthmann, p. 17.
100. Biggs, p. 70; I have made modifications in the translation.
101. W. G. Lambert, "Devotion: The Languages of Religion and Love," in *Figurative Language in the Ancient Near East* (London: University of London, 1987), pp. 30–31; Thompson, *DAB*, p. 255, says it may be "gall-apple" from oak trees (cf. p. 304), but, as noted earlier (n. 71) Thompson's identifications are not definitive.
102. Lambert, "Devotion" (1987), p. 27; P. Attinger, "Enki et Ninhursaĝz," *Zeitschrift für Assyriologie und vorderasiatische Archaeologie* 74 (1984): 20–22, lines 148–150; 22, lines 175–177; for cucumber or apricots [*U'KUŠ*]; apples [*HAŠHUR*], grapes [*GEŠTIN*].
103. Ibid., p. 28; F. Thureau-Dangin, "Un Hymne à Išta de lat haute Époque Babylonienne," *Revue d'assyriologie et d'archéologie* 22 (1925):172, 174, lines 7–8 in French trans.
104. Ibid., p. 29; Walter Farber, *Beschwörungsrituale an Istar und Dumuzi* (Wiesbaden: Steiner, 1977), p. 185, cf. 202.
105. Kramer, *The Sacred Marriage Rite*, p. 57; Zainab Bahrani, "Performativity and the Image: Narrative, Repesentation, and the Uruk Vase," in *Leaving No Stones Unturned: Essays on the Ancient Near East and Egypt in Honor of Donald P. Hansen*. Erica Ehrenberg, ed. (Winona Lake, IN: Eisenbrauns, 2002), pp. 15–22.
106. "The Joy of Sumer: The Sacred Marriage Rite," in *Inanna*, pp. 107–108.
107. From poem "The Farmer of Enlil," in Kramer, *Sacred Marriage*, p. 54.
108. J. S. Cooper, "Sacred Marriage and Popular Cult in Early Mesopotamia," in *Official Cult and Popular Religion in the Ancient Near East*, E. Matsushima, ed. (Heidelberg: Universitätsverlag C. Winter, 1993), p. 83 [81–96]; accepting the Uruk vase as depicting the Sacred Marriage is Zainab Bahrani, *Women of Babylon: Gender and Representation in Mesopotamia* (London and New York: Routledge, 2001), pp. 134–140.

109. W. W. Hallo, "The Birth of Kings," in *Love and Death in the Ancient Near East: Essays in Honor of Marvin H. Pope*. J. H. Marks and R. M. Good, eds. (Guildford, CT: Four Quarters, 1987), pp. 48–49 [45–52].
110. Balz-Cochois, p. 47.
111. Wolkstein and Kramer, pp. 16–17.
112. Ibid., p. 26.
113. *Epic of Gilgamesh,* N. K. Sandars, trans. (Harmondsworth, NY: Penguin Books, 1972), p. 63.
114. Alexander Heidel, *Gilgamesh Epic and Old Testament Parallels* (Chicago: University of Chicago Press, 1963), pp. 27–33, and Speiser in Pritchard, *ANET,* pp. 74–75, 77–78. There are even some who have suggested that the Ezekiel 28: 12–15 account of Adam may be drawn from an independent Mesopotamian account, but in it the first man while in Eden is described as "blameless in your ways from the day you were created till iniquity was found in you" (15). Morris Jastrow, *Religion of Babylonia and Assyria* (New York: Grinn and Co., 1898), p. 475–487.
115. *Epic of Gilgamesh,* Sandars trans. pp. 64–66.
116. Heidel, *Gilgamesh*, pp. 27–33, and Speiser in Pritchard, *ANET,* pp. 74–75, 77–78.
117. Bleibtreu, p. 76; Börker-Klähn in *RA* 3: 620.
118. Wolkstein and Kramer, pp. 9, 180–181; Geo Widengren, *The King and Tree of Life in Ancient Near Eastern Religion* (Uppsala: A-B. Lundequistskea Bokhandeln, 1951).
119. Adolf Butenandt and H. Jacobi, "Über die Darstellung eines krystallisierten pflanzliche Tokokinins (Thelykinins) und seine Identifizierung mit dem α-Follikelhormon," *Zeitschrift für physiologische Chemie*, 218(1933): 104–112.
120. Ebers 799 in Hildegard von Deines, Hermann Grapow, and Wolfhart Westendorf, *Grundriss der Medizin der alten Ägypter*, 9 vols. (Berlin: Akademie Verlag, 1954–1973), 4, pt. 1, pp. 279–280; pt. 2, pp. 211–212.
121. See references in John M. Riddle, *Contraception and Abortion from the Ancient World to the Renaissance* (Cambridge, MA: Harvard University Press, 1992), pp. 19–20.
122. Talmud, *Sabbath,* 110a (Epstein ed., 1, pp. 533–537).
123. Thompson, *DAB*, p. 289 (***GIŠ.KAL***= *ú-šu-ú*, "probably a willow"). This identification is based in part on Bruno Meissner, *Assyriologische Studien* VI (= *Mitteilungen der vorderasistischen/aegyptischen Gesellschaft* 18/II. Berlin, 1913), 1114. 2. The rendering of *ušû* as "willow" is subject to considerable debate. See W. von Soden, *Akkadisches Handwörterbuch* (Wiesbaden: Otto Harrassowitz, 1981), 1442b. R. Campbell Thompson's pioneering efforts to identify plants in cuneiform script is subject to revisions by modern philologists whose skills are more advanced.
124. *DAB*, p. 289.
125. *DAB*, p. 291. See Albert T. Clay, *Documents from the Temple Archives of Nippur (=Babylonian Expedition of the University of Pennsylvania)* Series A: Cuneiform Texts. Philadelphia, 1906) vol. 14, no. 163, 48.
126. *DAB*, p. 291.
127. Ibid.
128. Dioscorides, *De materia medica,* 1. 104 (1. 4, Max Wellmann ed., Berlin, 1958 repr., vol. 1, p. 96).
129. Soranus, *Gynecology.* Oswei Temkin, trans. (Baltimore: Johns Hopkins University Press, 1956), 1. 65; elsewhere see Riddle, *Contraception*, pp. 47–48, 56, 98, 122–123, 161.

130. See Mott T. Greene, *Natural Knowledge in Preclassical Antiquity* (Baltimore: Johns Hopkins University Press, 1992), p. 125–137; Mary A. K. Matossian, *Poisons of the Past* (New Haven, CT: Yale University Press, 1989), pp. 9–14, 57.
131. Robert D. Biggs, "Ergotism and Other Mycotoxicoses in Ancient Mesopotamia," *Aula Orientalis. Revista de estudios del Próximo Oriente Antiguo* 11 (1991): 20 [15–21].
132. Ibid.
133. Marie Delcourt, *Stérilité mystérieuse & naissances maléfiques dans l'antiquité classique* (Liège: Faculté de Philosophie et Lettres, 1938).
134. Elaine Adler Goodfriend, "Prostitution (OT)," *ABD,* 5:505–510.
135. *Atra-hasī. The Babylonia Story of the Flood.* W. G. Lambert and A. R. Millard, trans. with *The Sumerian Flood Story,* M. Civil trans. (Oxford: Clarendon Press, 1969), p. 109, line 39 of *Atra-hasī* myth.
136. Ibid., Tablet BE39099, reverse ii, lines 12–13.
137. Anne Draffkown Kilmer, "The Mesopotamian Concept of Overpopulations and Its Solution as Reflected in the Mythology," *Orientalia* n.s. 41 (1972): 160–177.
138. Some of the Akkadien prescriptions for women are reproduced, transcribed and translated by René Labat, *Traité akkadien de diagnostics et pronostics médicaux,* 2 vols. (Paris: Academie Internationale d' Historire des Sciences, 1951): 1: 200–217, and by Köcher, *MTU* 3: 19 (without translation).
139. Herodotus, 1. 195 cf. evaluation by Terry R. Glover, *Herodotus* (Freeport, NY: Books for Libraries Press, 1969 repr.), p. 68; see also *The Cambridge History of Classical Literature,* 2 vols. (Cambridge: Cambridge University Press, 1985) 1: 416–441.
140. Mylitta is presumed to be synonymous with Inanna and Isthar, as the goddess of flowing water, fertility, and sexuality.
141. Herodotus, 1. 199 (A. D. Godley, trans., in Loeb: Cambridge, MA: Harvard University Press, 1990 repr., 1: 251–253).
142. Herodotus, 1. 199 (Godley, 1985, 1: 253).
143. Julia Assante, "From Whore to Hierodules: The Historiographic Invention of Mesopotamian Female Sex Professionals," in *Ancient Art and Its Historiography.* A. A. Donohue and Mark D. Fullerton, eds. (Cambridge: Cambridge University Press, 2003), pp. 13–68,
144. Ibid., pp. 32, 34.
145. I have found a French study, albeit an older study, to be comprehensive in evidence presented: Paul Lacroix, *A History of Prostitution.* Samuel Putnam, trans. from French (New York: Covici, Friede Publishers, 1931).
146. Stephanie Lynn Budin, *The Myth of Sacred Prostitution in Antiquity* (New York: Cambridge University Press, 2008), p. 1.
147. Vinciane Pirenne-Delforge, "Review of: Stephanie Budin, *The Myth of Sacred Prostitution...,"* in *Bryn Mawr Classical Review* (2009). 04. 28 online review. p. 8.
148. Budin, pp. 58–89, quoting, p. 71.
149. Pirenne-Delforge, p. 3.
150. Bottéro (2001), pp. 90–126.
151. Louis Delaporte, *Mesopotamia: The Babylonian and Assyrian Civilization.* V. Gordon Childe, trans. (New York: Barnes and Noble, repr. 1970), p. 284 (on family life, pp. 281–290).
152. William W. Hallo, "The Slander Bride," in *Studies Presented to A. Leo Oppenheim, June 7, 1964* (Chicago: University of Chicago Press, 1964), pp. 95–105.
153. Hector Avalos, *Illness and Health Care in the Ancient Near East: The Role of the Temple in Greece, Mesopotamia, and Israel* (Atlanta: Scholars Press, 1995), pp. 28–30;

A. Leo Oppenheim (*Ancient Mesopotamia: Portrait of a Dead Civilization* [Chicago: University of Chicago Press, 1964], p. 106) cautions that, although we have abundant details of lower temple personnel, the institution of the Mesopotamian temple "is very much in the dark."

154. Ibid., pp. 101–131, esp. 107.
155. C. J. Mullo-Weir, "Four Hymns to Gula," *Journal of the Royal Asiatic Society* (January 1929): 16–17, lines 25–29.
156. W. G. Lambert, "The Gula Hymn of Bullutsa-rabi," *Orientalia* 36 (1967): 121, lines 79–83.
157. Avalos, pp. 172–182.
158. Stoll, p. 114. On priestesses acting as midwives, see Stoll, p. 112; slightly different translation in *Atra-Hasī,* Lambert-Millard trans. lines 15–19, p. 63.
159. Ibid., p. 193.
160. Ibid., pp. 196–202.
161. Marc Van De Mieroop, "Women in the Economy of Sumer," in *Women's Earliest Records from Ancient Egypt and Western Asia.* Barbara S. Lesko, ed. (Atlanta: Scholars Press, 1989), p. 54. Van De Mieroop was asked whether women lived in attached houses around temples, in dormitories, or in villages outside the temple's walls; he said, "Well, we don't really have information about that, especially not archaeological information, but it is my opinion that these dependent laborers did have a family life and probably did own a house or were assigned a house"... where they worked and raised their children. (p. 67).
162. Bahrani (2001), p. 122.
163. Martha T. Roth, "Marriage and Matrimonial Prestations in First Millennium B.C. Babylonia," in *Women's Earliest Records*, p. 246.
164. Delaporte, p. 311.
165. See opinion of J. J. Bachofen, *Myth, Religion and Mother Right.* Ralph Manheim, trans. (Princeton, NJ: Princeton University Press, 1967), pp. 150–151.
166. Leonard W. King, *A History of Sumer and Akkad* (London: Chatto & Windus, 1916), p. 90.
167. See Chapter 1, "The First Schools," in *History Begins at Sumer: Thirty-nine Firsts in Man's Recorded History.* Samuel Noah Kramer ed. 3rd ed. (Philadelphia: University of Pennsylvania Press, 1981), pp. 3–9.
168. Rivkah Harris, "The *Nadītu* Woman," in *Studies Presented to A. Leo Oppenheim, June 4, 1964* (Chicago: The Oriental Institute, 1964), pp. 106–109.
169. Merlin Stone, *When God Was Woman* (New York: Dial Press, 1976), p. 40; Johanna H. Stuckey, "Sacred Prostitutes," *MatriFocus: Cross-Quarterly for Goddess Woman Samhain* (Vol. 5–1): http:www.matrifocus.com/SAM05/spotlight.htm (accessed 4/6/09).
170. Allan Zagarell, "Responses to Prof. Kuhrt's Paper," in *Women's Earliest Records*, p. 241; Bernard Frank Batto, *Studies on Women at Mari* (Baltimore: Johns Hopkins University Press, 1974), pp. 79–92, says that the equivalent priestess at Mari, called *ughabtum*, lived either in a cloister or in the home of her father, unmarried, and remained childless.
171. Beatrice A. Brooks, "Fertility Cult Functionaries in the Old Testament," *Journal of Biblical Literature* 60/3 (1941): 231.
172. John A. Halloran, Sumerian Lexicon: http://www.sumerian.org/suml-r.htm (accessed 4/6/09).
173. Halloran: http://www.sumerian.org/sumg-k.htm (accessed 4/6/09); Budin, pp. 20–33.
174. Stone, p. 40; see discussion of *naditu* by Budin, pp. 22–23.

175. Harriet Crawford, *Sumer and the Sumerians,* 2nd ed. (Cambridge: Cambridge University Press, 2004), esp. pp. 89–114; Bahrani (2001), pp. 51–55, cautions against a tendency to ascribe cultic prostitution to any figurine or relief that depicts sexual intercourse.
176. Thornkild Jacobsen, "Pictures and Pictorial Language (The Burney Relief)," in *Figurative Language in the Ancient Near East* (London: School of Oriental and African Studies, 1987), pp. 5–6.
177. Ibid.
178. Jean Bottéro, *Mesopotamia, Writing, Reasoning, and the Gods.* Zainab Bahrani and Marc Van De Mieroop trans. (Chicago: University of Chicago Press, 1992; French ed. 1987), p. 190.
179. Ibid., pp. 189–190.
180. Karel van der Toorn, "Prostitution (Cultic)," *ABD,* 5:510–513, with summary of scholarship and references; see also Elaine Adler Goodfriend, "Prostitution (OT)," *ABD,* 5:505–510.
181. R. Harris, "Independent Women in Ancient Mesopotamia?" in *Women's Earliest Records,* pp. 145–165. *Proceedings of the Conference on Women in the Ancient Near East, Brown University, Providence, Rhode Island, November 5–7, 1987*; cf. Julia Assante, "From Whore to Hierodules: The Historiographic Invention of Mesopotamian Female Sex Professionals," in *Ancient Art and Its Historiography.* A. A. Donohue and Mark D. Fullerton, eds. (Cambridge: Cambridge University Press, 2003), pp. 13–68 (esp. p. 25).
182. Stoll, p. 38; Brian Lewis, *The Sargon Legend: A Study of the Akkadian Text and Tale of the Hero Who Was Exposed at Birth* (Cambridge, MA: American School of Oriental Research, 1980), pp. 24, 44.
183. Thompson, *DAB,* p. 66; John M. Riddle, *Eve's Herbs: A History of Contraception and Abortion in the West* (Cambridge, MA: Harvard University Press, 1997), p. 55.
184. Procopius, *Anecdota,* 9. 19 (G. A. Williamson, trans., Penguin, 1966).

CHAPTER 2

1. Elizabeth Gould Davis, *The First Sex* (New York: Putnam, 1971), p. 142; Robert Graves, *Adam's Rib and Other Anomalous Elements in the Hebrew Creation Myth* (New York: Thomas Yoseloff, 1958), p. 7.
2. Ronald S. Hendel, "Genesis, Book of," in *ABD,* 2: 938–941 (with extensive bibliography); dated but useful is John Skinner, *Critical and Exegetical Commentary on Genesis,* 2nd ed. (Edinburgh: T. & T. Clark, 1930), 2: 57–97; Hermann Gunkel, *Das Buch Genesis* (*GHK*; Göttingen, 1922); John L. McKenzie, "The Literary Characteristics of Genesis 2–3," *Theological Studies* 15 (1954): 545–546 [541–572]; Robert Graves and Raphael Patai, *Hebrew Myths: The Book of Genesis* (Garden City, NY: Doubleday, 1964), esp. pp. 60–81.
3. "Myth of Adapa," Tablet 1.4 in Robert Williams Rogers, *Cuneiform Parallels to the Old Testament* (New York: Abingdon Press, 1912), p. 69; also via Web site: http://www.sacred-texts.com/ane/adapa.htm (accessed 5/5/08).
4. John A. Phillips, *Eve: The History of an Idea* (San Francisco: Harper & Row, 1984), p. 3.
5. Howard N. Wallace, "Eve," in *ABD,* 2:676–677.
6. Pamela Norris, *Eve: A Biography* (New York: New York University Press, 1999), p. 19.
7. Ibid.

8. Joan O'Brien, "Nammu, Mani, Eve and Pandora: 'What's in a Name?'" *Classical Journal* 79/1 (1983): 38.
9. On derivation of the name Adam, see Howard N. Wallace, "Adam" *ABD,* 1:62–64.
10. E. A. Speiser, *The Anchor Bible Genesis* (Garden City, NY: Doubleday, 1964), 1: 26.
11. J. Coppens, *La connaissance du bien et du mal et le péché du paradis* (Gembloux, 1948), pp. 26–28, 99–134.
12. Jeffrey Burton Russell, *Satan: The Early Christian Tradition* (Ithaca, NY: Cornell University Press, 1981).
13. *Life of Adam and Eve* (M. D. Johnson, trans.) in J. H. Charlesworth, ed. *Old Testament Pseudoepigrapha,* 2 vols. (Garden City, NY: Doubleday, 1985), 2: 249–295.
14. John J. Scullion, "Genesis, The Narrative of," in *ABD,* 2: 945.
15. Claus Westermann, *Genesis 1–11. A Commentary.* John J. Scullion, trans. (Minneapolis: Augsburg, 1984 [English ed; 1974 German ed.]), pp. 237–242.
16. Karen Randolf Joines, *Serpent Symbolism in the Old Testament* (Haddonfield, NJ: Haddonfield House, 1974), pp. 64–68.
17. Wallace, *ABD,* 2: 677.
18. Joines, p. 18.
19. Tablet XI, lines 279–281.
20. Joines, pp. 16–31.
21. *The Dictionary of World Myth,* Peter Bently, ed. (New York: Facts on File, 1995), p. 196; see documentation provided by Coppens, pp. 99–134, as serpent as fertility symbol in ancient societies.
22. John L. McKenzie, "The Literary Characteristics of Genesis 2–3," *Theological Studies* 15 (1954): 564.
23. Ambrose, *Hexameron, Paradise.*13. 61–62, John J. Savage, trans. (New York: Fathers of the Church, 1961), p. 343.
24. Börker-Klähn in *RA* 3:620–621.
25. Geo Widengren, *The King and Tree of Life in Ancient Near Eastern Religion* (Uppsala: A-B. Lundequistskea Bokhandeln, 1951), pp. 38–39; cf. Pritchard, *ANET,* p. 96, lines 286–296.
26. J. Andrew McDonald, "Botanical Determination of the Middle Eastern Tree of Life," *Economic Botany* 56/2 (2002): 113–129.
27. James B. Pritchard, *The Ancient Near East: An Anthology of Text and Pictures* (Princeton, NJ: Princeton University Press, 1958), 1: 74.
28. Widengren, pp. 7, 21; *Interpreter's Dictionary of the Bible* 3: 840.
29. *Epic of Gilgamesh.* N. K. Sandars, trans. (Penguin rev. ed., 1972), p. 116; slightly different trans. by J. B. Pritchard, *Ancient Near Eastern Texts Related to the Old Testament,* 3rd ed. (Princeton, NJ: Princeton University Press, 1972), p. 96, lines 278–288. (hereinafter cited as *ANET*).
30. Ibid., p. 102; see similar translation in James B. Pritchard, *The Ancient Near East: An Anthology of Text and Pictures,* 2nd ed. (Princeton, NJ: Princeton University Press, 1958), p. 64.
31. Pritchard, *ANET.*
32. Howard N. Wallace, "Tree of Knowledge and Tree of Life," in ABD, 6: 657.
33. Elaine Pagels, *Adam, Eve, and the Serpent* (New York: Random House, 1988), p. 27.
34. John Skinner, *A Critical and Exegetical Commentary on Genesis,* 2nd ed. (Edinburgh: T. & T. Clark, 1930), p. 94.
35. 2Sam. 14:17; 19:35; Deut. 1:39; Isa. 7:15–16.
36. J. M. Evans, *Paradise Lost and the Genesis Tradition* (Oxford: Clarendon Press, 1968), p. 18.

37. *Apocryphal Lives of Adam and Eve*. 72. Brian Murdoch and J. A. Tasioulas, eds. (Exeter: University of Exeter Press, 2002), p. 40.
38. Immanuel Löw, *Die Flora der Juden* (Hildesheim: Georg Olms, 1967), 3: 212–238, believes, however, that the Genesis reference is to some kind of apple.
39. *Interlinear Hebrew/Greek English Bible*, 4 vols. Jay Green, ed. and trans. (Wilmington, DE: Associated Publishers and Authors, 1976), 1:6.3.2.
40. *The Septuagint with Apocrypha: Greek and English*. Sir Lancelot C. Brenton, ed. (Grand Rapids, Michigan: Zondervan repr. of 1851 ed.), pp. 3–4 for Genesis 2: 3–4.
41. Zohary and Hopf, p. 174.
42. Löw, 3:212–238.
43. Wildred G. Lambert, "Devotion: The Languages of Religion and Love," in *Figurative Language in the Ancient Near East* (London: University of London, 1987), pp. 30–31.
44. Irene and Walter Jacob, "Flora," *ABD*, 2:806–807.
45. J. D. Douglas, "Apple," *The New Bible Dictionary* (Grand Rapids, Michigan: Eerdmans, 1962), pp. 50–51; M. A. Powell, "Classical Sources and the Problem of the Apricot," *Bulletin on Sumerian Agriculture* 3 (1987): 156 [153–156], proposes that the Sumerian term *HAŠHUR* has a "minimal probability of being correct" as being an apricot. Powell (p. 154) agrees with those who reject the Genesis *tpwh*. as being an apricot.
46. Proposed by Friedrich Muthmann, *Der Granatapfel Symbol des Lebens in der alten Welt* (Bern: Abegg-Stiftung, 1982), pp. 14, 16 and Löw, 4:5543; interestingly the pomegranate was not indigenous to Egypt, having come there during the New Kingdom, the period, of course, when the Israelites are said to be there before the Exodus. See Lise Manniche, *An Ancient Egyptian Herbal* (Austin: University of Texas Press, 1989), p. 139, and discussion later in this chapter.
47. Exodus 28: 33–35; cf. 39: 24–26; Josephus (*Antiquities of the Jews*. 3. 160 [7. 4]) repeated the description, but in *Wars of the Jews* (5. 231–232 [5. 7]). *The Interpreter's Bible*, 12 vols. (New York: Abingdon Press, 1952) 1: 1043 gives a decidedly nonhistoric explanation: "Bells and pomegranates, a sweet sound and a sweet savor."
48. Philo, *De vitae Mosis* in *The Works of Philo: Complete and Unabridged*. C. D. Yonge, trans. (Peabody, MA: Hendrickson, 1993), 2. 24. 119–121, who said that pomegranates symbolize the flowing of water and bells for concord and harmony.
49. *ABD*, 5: 808.
50. Josephus, *Antiquities of the Jews*. 3. 160 [7. 4]; *Wars of the Jews*. 5. 231–232 [5. 7].
51. Paul Jacobsthal, *Greek Pins* (Oxford: Clarendon Press, 1956), p. 189.
52. Widengren, pp. 38–39.
53. Harold N. Moldenke and Alma L. Moldenke, *Plants of the Bible* (Waltham, MA: Chronica Botanica Company, 1952), pp. 189–191, provide a complete listing of biblical references to the pomegranate and connect them with fertility and from earliest times as being sacred.
54. I Kings 7: 15–22, 41–42; 2 Kings 25:17; 2 Chronicles 3: 15–17; Jeremiah 52: 21–23. *The Interpreter's Bible* (3: 63) explains the pomegranates presence as "emblems of fertility because of the extraordinary number of pips which they contain."
55. R. B. Y. Scott, "The Pillars Jachin and Boaz," *Journal of Biblical Literature* 58/2 (1939): 145; Carol Meyers, "Jachin and Boaz," in *ABD*, 3:597–598; Carol Meyers, "Jachin and Boaz," *Catholic Biblical Quarterly* 45/2 (1983): 167–178.
56. *Interpreter's Bible* (1954) 3: 63.
57. "Update. Finds or Fakes; Ivory Pomegranate," *Biblical Archaeology Review* 31/2 (2005): 62–63; see Hershel Shanks, "Pomegranate Sole Relic From Solomon's

Temple, Smuggled out of Israel, Now Recovered," *Moment. Magazine of Jewish Culture and Opinion* (December 1988), p. 41. For his research on the vase, I am indebted to Robert West's Honors Thesis, "House of the Lord Vase," (unpubl., December, 2006).

58. André Lemaire, "Probable Head of Priestly Scepter from Solomon's Temple in Jerusalem," *Biblical Archaeology Review* 10 (1984): 27.
59. "Update. Finds or Fakes; Ivory Pomegranate," *Biblical Archaeology Review* 31/2 (2005): 62–63; (see Shanks, p. 41).
60. Nahman Avigad, "The Inscribed Pomegranate from the 'House of the Lord,'" *The Biblical Archaeologist* 53/3 (1990): 157–166.
61. Lemaire, pp. 27–28.
62. Kathleen M. Kenyon, *Excavations of Jericho*, 2 vols. (London: British School of Archaeology in Jerusalem, 1965), 1:392–393; 2: 465.
63. Navigad, p. 162.
64. David Hendin, *Guide to Ancient Jewish Coins* (New York: Attic Books, 1976), p. 16.
65. Ibid., p. 16.
66. See image: http://www.remnantofyhwh.com/images/Jewish5.jpg (accessed 5/14/09). I suspect that the three pomegranates stood for the three Israeli offices of king, priest, and prophet.
67. http://ferrelljenkins.wordpress.com/2008/02/11/israel-issues-a-new-2-shekel-coin/ (accessed 6/8/08).
68. Sara A. Immerwahr, "The Pomegranate Vase: Its Origins and Continuity," *Hesperia* 51/4 (1989): 397–410.
69. Ibid., pp. 399–400.
70. Ibid., pp. 40–41; in Staatliche Museum, Berlin, Antique Collection.
71. Hugh G. Evelyn-White's good translation is available online from 1914 Loeb Classical Library edition. translated by Hugh G.Evelyn-White http://www.sacred-texts.com/cla/demeter.htm (accessed 5/15/08).
72. Ann Suter, *Narcissus and the Pomegranate: An Archaeology of the Homeric Hymn to Demeter* (Ann Arbor: University of Michigan Press, 2002), p. 5.
73. Helene P. Foley, *The Homeric Hymn to Demeter* (Princeton, NJ: Princeton University Press, 1994), p. 107.
74. Ovid, *Metamorphoses,* 5. 534–542, for seven; *Fasti* 4. 607–608, for three; Foley, *Homeric Hymn to Demeter*, pp. 56–57.
75. Suter, p. 90.
76. Ibid.
77. *Hymn to Demeter,* 411–413 (Helene P. Foley); John L. Myres, "Persephone and the Pomegranate (*H. Dem.* 372–374)," *Classical Review*, 52/2 (1938): 51–52, sees the pomegranate as a love-charm.
78. Ibn Sīnā, *Liber canonis.* ii. 2. 578 (translated from Arabic by Meyerhof, in Himes 1936, pp. 142–143; Latin translation by Gerard of Cremona, 1507, fol. 146v).
79. Foley, pp. 56–57; 108.
80. Claudian. *De raptu Proserpinae* 1. 28–36, 98–110 (David Slavitt, trans.); see also discussion by Stephen M. Wheeler, "The Underworld Opening of Claudian's *De Raptu Proserpinae*," *Transactions of the American Philological Association* 125 (1995): 113–134, esp. 113, and Foley, pp. 108–112.
81. See interpretations by Wheeler, pp. 113, 128–129.
82. Suter, p. 48.
83. Aristophanes, *Thesmophoriazusae,* pp.390–395 (Benjamin B. Rogers, trans., p. 165).

84. J. Prytz Johansen, "The Thesmophora as a Women's Festival," *Temenos* 11 (1975): 80.
85. Ibid., p. 84.
86. Ibid., p. 85.
87. Hesiod, *Work and Days,* 504, H. G. Evelyn-White, trans. Classical Electronic Texts: http://www.theoi.com/Text/HesiodWorksDays.html (accessed 4/26/08); *Hesiod,* 2 vols. (Cambridge, MA: Harvard University Press, 2006–2007), 1: 128–131.
88. Ovid. *Metamorphoses* 5. 534, for seven; *Fasti* 4. 607–608, for three; *Hymn to Demeter.* 371–374, 411–413, unspecified number.
89. Claudian, pp. 406–408.
90. Athens Nat. Mus. 484 and reproduced by Jane Harrison, *Prolegomena to the Study of Greek Religion* (London: Merlin, 1961), p. 275. Harrison (p. 274) believes that Demeter and Persephone are combined in one representation.
91. For numerous other examples of the use of pomegranate in association with fertility, see Muthmann, *Granatapfel.*
92. Soranus, *Gynecology.* Oswei Temkin, trans. (Baltimore: Johns Hopkins University Press, 1956), 1. 62 (pp. 64–65).
93. See John M. Riddle, *Contraception and Abortion from the Ancient World to the Renaissance* (Cambridge, MA: Harvard University Press, 1992), pp. 25–26, 32–33, 51–53, 57, 88–89, 94–97, 102, 120, 126, and 132; not useful is Warren R. Dawson, "Studies in Ancient Materia Medica: The Pomegranate," *American Druggist,* December 1925, pp. 22–24.
94. Riddle (1992), pp. 82–92.
95. Ibid., pp. 102, 120, 126, and 132.
96. David Zohary and Maria Hopf, *Domestication of Plants in the Old World.* 3rd ed. (Oxford: Oxford University Press, 2000), pp. 170–171; Mary Anne Murray, "Fruits, Vegetables, Pulses and Condiments," in *Ancient Egyptian Materials and Technology.* Paul T. Nicholson and Ian Shaw, eds. (Cambridge and New York: Cambridge University Press, 2000), pp. 624–625.
97. Juvenal, *Satirae,* 6. 595 (Green trans. p. 149); see also John M. Riddle and J. Worth Estes, "Oral Contraceptives in Ancient and Medieval Times," *American Scientist* 80 (1992): 226–233, and Riddle, "Coins and Contraceptives: The Plant That Made Kyrene Famous," *Celator* 17/12 (2003): 34–35.
98. Manniche, p. 139; Alfred Lucas and J. R. Harris, *Ancient Egyptian Materials and Industries,* 4th ed. (London: Arnold, 1962), p. 36; for other authorities, see Immerwahr, p. 402 fn.
99. F. N. Hepper, *Pharaoh's Flowers: The Botanical Treasures of Tutankhamun* (London: Her Majesty's Stationary Office, 1990), p. 64.
100. Immerwahr, pp. 400–403.
101. Hepper, p. 62.
102. Riddle (1992), pp. 66–73.
103. Manniche, p. 140 citing Ebers 50 and 63; see also Cyril P. Bryan, *Ancient Egyptian Medicine: The Papyrus Ebers* (Chicago: Ares, 1974), pp. 120–121.
104. Dioscorides, *De materia medica.* 1. 108 (Lily Y. Beck, trans. 79).
105. Ibid., 1. 108 and 109.
106. I recall seeing a terra cotta vase, dated Middle Kingdom (11th Dynasty), in the Cairo Museum, that had pomegranates as borders, but I have been unable to find a reference to the vase.
107. Hippocratic Corpus, *Aphorisms* 1 (J. Chadwick and W. N. Mann, trans., *Hippocratic Writings.* [Penguin pb., 1978], p. 148).

108. *Yebamoth.* 8.4 (Babylonia Talmud).
109. Fred Rosner, *Medicine in the Bible and the Talmud* (Yeshiva University Press, 1995), p. 93; David M. Feldman, *Birth Control in Jewish Law* (New York: New York University Press, 1968), p. 38.
110. John Riddle, "Women's Medicines in Ancient Jewish Sources: Fertility Enhancers and Inhibiters," in *Disease in Babylonia.* I. L. Finkel and M. J. Geller, eds. (Boston: Leiden, 2007), p. 202.
111. Ibid., pp. 206–210.
112. Merlin Stone, *When God Was a Woman* (New York: Dial Press, 1976), p. 217.
113. Evans, p. 19.
114. Muthmann, p. 51; now in Würzburg, Matin von Wagner Museum der Universität Würzburg.
115. Ibid., pp. 102–103.
116. Now in London, British Museum; J. M. C. Toynbee, "A New Roman Mosaic Pavement Found in Dorset," *Journal of Roman Studies* 54 (1964): plate 2; fig. 1, p. 14 for discussion. Also Muthmann, pp. 114–115; Börker-Klähn, "Granatapfel," *RA* 3: 627.
117. James Hall, *Dictionary of Subjects and Symbols in Art.* Rev. ed. (New York: Harper and Row, 1974), p. 249.
118. M. L. Handley, "Quercia," *Catholic Encyclopedia* 1911 ed. 12: 601.
119. Hall, p. 249.
120. Börker-Klähn, in *RA*, 3: 627.
121. Philips, p. 29; cf. similar observations by Norris, pp. 40–58.
122. Norris, p. 65.
123. Philips, p. 45.

CHAPTER 3

1. *Midrash Rabbah,* 72. 5, in 5 vols. H. Freedman and Maurice Simon, trans. (London: Soncino Press, 1977), 1: 665.
2. Louis Ginzberg, *The Legends of the Jews,* 7 vols. (Philadelphia, PA: Jewish Publication Society of America, 1937), 1: 367.
3. Ibid.
4. Jewish Encyclopedia online but be warned: naming the twelve tribes is a confusing task: http://www.jewishencyclopedia.com/view.jsp?artid=326&letter=T&search=Twelve%20tribes (accessed 6/24/08).
5. John Skinner, *A Critical and Exegetical Commentary on Genesis,* 2nd ed. (Edinburgh: T. & T. Clark, 1976 repr.), p. 388; see also Skinner, *Myths and Legends of Flowers, Trees, Fruits, and Plants in All Ages and All Climes* (Philadelphia, PA: J. B. Lippincott, 1925), pp. 168–170.
6. J. K. Rowling, *Harry Potter and the Chamber of Secrets* (New York: Scholastics, 1999), pp. 91–93.
7. Qu'ran. Surah 14:37, cf. Gen. 21:14; Surah 11:69–72; Surah 37:100–112. For a full discussion, see Phyllis Tribble and Letty M. Russell, eds. *Hagar, Sarah, and Their Children* (Louisville, KY: Westminster John Knox Press, 2006).
8. Jacob Neusner, *Comparative Midrash: The Plan and Program of Genesis and Leviticus Rabbah* (Atlanta: Scholars Press, 1986), p. 156; the importance of fertility and conception in ancient West Asia is explored by Gwendolyn Leick, *Sex and Eroticism in Mesopotamian Literature* (London: Routledge, 1994), pp. 45–48.

9. Adele Reinhartz and Miriam-Simma Walfish, "Conflict and Coexistence in Jewish Interpretation," in *Hagar, Sarah, and Their Children*, pp. 107–110.
10. Raphael Patai, *Sex and Family in the Bible and the Middle East* (Garden City, NY: Putnam's Sons, 1959), p. 121.
11. *Midrash Rabbah.* 71. 6. H. Freedman and Maurice Simon, trans., 5 vols. (London: Soncino Press, 1977), 1: 657.
12. Ibid., p. 77.
13. Savina J. Teubal, *Sarah the Priestess: The First Matriarch of Genesis* (Athens, OH: Swallow Press, 1984), p. 102; for a good discussion of the Genesis passage in reference to Jacob, see http://www.israelect.com/reference/WillieMartin/The_Story_Of_Judah.htm (accessed 6/14/08); Teubal pp. 71–72, proposes an interesting theory that Sarah, Rebekah, and Rachel were "associated in an important way with a religious order." Teubal connects Sarah's role as similar to Inanna's standing in front of the storehouse door as shown on the Uruk vase. The priestess order is the reason that Rachel, not Leah, is the protagonist in the Genesis story. Teubal (pp. 102–103, 135–136) believes that Sarah was the goddess "incarnate" in the mode of Inanna, Abraham was the "intermediary," and the narrative indicates a transition from a matriarchal to the patriarchal society.
14. Betty P. Jackson and Michael I. Berry, "Hydroxytropane Tiglates in the Roots of *Mandragora* Species," *Phytochemistry* 12 (1973): 1165.
15. M. Grieve, *A Modern Herbal* (New York: Barnes and Noble, repr. 1996), pp. 510–511.
16. Gaston Maspero, *A History of Egypt, Chaldea, Syria, Babylonia, and Assyria.* 13 vols. (London: Gralier Society [1901?–1906]), 1: 235; with some details found in Ernest Alfred Wallis Budge, *The Gods of the Egyptians or Studies in Egyptian Mythology*, 2 vols. (London: Methuen, 1904), 1: 363–366.
17. Ibid., 1:235–237; Lewis Spence, *Myths and Legends of Ancient Egypt* (New York: Farrar and Reinhart, 1933), pp. 162–168.
18. Richard H. Wilkinson, *Complete Gods and Goddesses of Ancient Egypt* (London: Thames and Hudson, 1988), pp. 138–145, esp. 141.
19. Ibid., pp. 184–186.
20. Ibid., p. 186.
21. C. J. S. Thompson, *The Mystic Mandrake* (New Hyde Park, NY: University Books, 1968), p. 44, who quotes (without citation) "Ebers" as saying that the mandrake beer from Elephantine was used as an intoxicating narcotic "in medicine and magic."
22. Manniche, pp. 117–119, esp. 119; see also Renate Germer, *Flora des pharaonichen Ägypten* (Mainz: Philipp von Zabern, 1985), pp. 42–43. John F. Nunn, *Ancient Egyptian Medicine* (New York: Cambridge University Press, 1996), pp. 156–157 accepts mandrake use in ancient Egyptian medicine but says there is "no evident that [powerful sedative properties] were known in the pharonic period." Whereas Nunn is correct, it defies reasonable judgment that it was employed but this pronounced quality was not perceived.
23. Carolyn Brown, "Plants and Trees of Ancient Egypt," *Inscriptions: The Newsletter of the Friends of Egypt Centre* 9 (December 2001), p. 11.
24. John M. Riddle, *Contraception and Abortion from the Ancient World to the Renaissance* (Cambridge, MA: Harvard University Press, 1992), pp. 66–69.
25. Kate Boasse-Griffithe, "The Fruit of the Mandrake," in *Fontes Aeque Pontes* (Wiesbaden: Otto Harrassowitz, 1983), pp. 62–72, has a full discussion of mandrake in the New Kingdom of Egypt.
26. F. Nigel Hepper, *Pharaoh's Flowers: The Botanical Treasures of Tutankhamun* (London: HMSO, 1990), p. 15.

27. Boasse-Griffithe, p. 76.
28. R. K. Harrison, "The Mandrake and the Ancient World," *The Evangelical Quarterly* 28/2 (1956): 56; see also Claudia Müller-Ebeling, "Die Alraune in der Bibel," in Alfred Schlosser, *Die Sage vom Galgenmännlein im Volksglauben und in der Literatur* (Berlin: Express Edition, 1987): 141–149.
29. Harold N. Moldenke and Alma L. Moldenke, *Plants of the Bible* (Waltham, MA: Chronica Botanica Company, 1952), p. 137; the Talmud tract Sanhedrin 99b two rabbis as identifying *duda'im* as *yabruchi* and *sabiski*, but according to Julius Preuss, *Biblical and Talmudic Medicine*, Fred Rosner, trans. (New York: Sanhedrin Press, 1978, based on German ed. of 1923), p. 463, both terms refer to the mandrake. A third rabbi says that it was *sigil*, a cypress; in the Midrash, more disagreement: one interpretation says that it was barley and another hackberry or myrtle berry (*mayish*). See *Midrash Rabbah*, 72, 2 (H. Freeman and Maurice Simon, trans., 5 vols. [London: Soncino Press, 1977], 1:662) but immediately in the next section (72. 3), the commentary assumes that mandrake is the correct identification. Preuss concluded by quoting a colleague (Huber) as saying that *duda'im* was highly problematical as being mandrake. Certainly in light of the Septuagint and Josephus (discussion below), the identification as mandrake is about as certain as we can come in interpreting old texts without sufficient context.
30. Löw, 2: 363–368; Preuss, p. 463; Michael Zohary, *Plants of the Bible* (Cambridge: Cambridge University Press, 1982), pp. 188–189; Moldenke and Moldenke, pp. 137–139, although the Moldenke team reject two proposals that the Genesis reference was to truffles (p. 283) and the American mandrake (ibid.).
31. Ibid., p. 648. For the German custom, Pope cites: "cf. *Handbuch des Deutschen Aberglaubens*, s.v. Mandragora" but I have been unable to find the reference.
32. *Ba'al et 'Anat*. 5.AB. C. 13, in *Textes Ougaritiques.*, André Caquot, Maurice Sznycer, and Andrée Herdner, trans with commentary (Paris, Les Éditions du Cerf, 1974,) 1: 164, but these translators contest the translation as pointing to the mandrake plant; a number of other Ugaritic scholars, however, believe the word does mean mandrake. See references in M. Stoll, *Birth in Babylonia and the Bible: Its Mediterranean Setting.* Cuneiform Monographs 14 (Groningen: Styx Publications, 2000), p. 56 fn.
33. Hesychius of Alexandria, *Hesychii Alexandini Lexikon*, 5 vols. (Amsterdam: Hakkert, 1965), 3: 69.
34. *Oxford English Dictionary,* electronic version, *s.v.* "mandragora."
35. Columela, *On Agriculture,* 3 vols. E. S. Forster and Edward H. Heffner trans. in Loeb Classical Library (Cambridge, MA: Harvard University Press; London: Heinemann, 1955, 1993 repr.), 10. 19.
36. Homer, *Odyssey,* 2 vols. A. T. Murray, trans. in Loeb Classical Library (Cambridge, MA: Harvard University Press, 1984), 10. 302–306.
37. Columela, 10. 20; Pliny, 25.94. 147.
38. Dioscorides, *De materia medica*, 2005, 4. 75.
39. Jerry Stannard, "The Plant Called Moly," in *Herbs and Herbalism in the Middle Ages and Renaissance.* Katherine Stannard and Richard Kay, eds. (Aldershot: Ashgate Variorum, 1999), pp. 256–307, esp. p. 281; on the other hand John Scarborough ("The Pharmacology of Sacred Plants, Herbs, and Roots," in *Magika Hiera: Ancient Greek Magic and Religion* [New York: Oxford University Press, 1991), pp. 139–140, suggests the possibility of "moly" being the opium poppy.
40. Homer, *Odyssey,* 10. 302—306 (A. T. Murray, trans. in Loeb, 1: 367.
41. See W. Gunther Plaut, Bernard J. Bamberger, and William H. Hallo, *Torah: The Torah, a Modern Commentary* (New York: Union of American Hebrew Congregation, 1981),

p. 207; Michael Wink, "A Short History of Alkaloids," in *Alkaloids. Biochemistry, Ecology, and Medicinal Aplications.* Margaret F. Roberts and Michael Wink, eds. (New York and London: Plenum Press, 1998), p. 21; R. L. Hunter, commentary to *Apollonius of Rhodes: Argonautica Book III* (Cambridge: Cambridge University Press, 1989), p. 188.

42. Anne Van Arsdall, "Exploring What Was Understood by 'Mandragora' in Anglo-Saxon England," in *Old Names, New Growth: Proceedings of the 2nd Anglo-Plant Name Survey Symposium.* Graz, June 6–10, 2007. P. Bierbaumer and H. W. Klug, eds. (Frankfurt/Main: Lang, 2009), p. 67. I express gratitude to Dr. Van Arsdall for permitting to see and quote from her prepublication paper. For detailed references to various legend accounts, see Charles Brewster Randolph, "The Mandrake of the Ancients in Folk-Lore and Medicine," *Proceedings of the American Academy of Arts and Sciences* 40 (1905): 487–537. C. J. S. Thompson's (1968), a book devoted to the mandrake is without reliable references. See also Wolfgang Schmidbauer, "Die magische Mandragora," *Antalos* 10 (1969): 274–286.
43. Theophrastus, *Enquiry into Plants,* 9. 7. 8 (Arthur Hort, trans., in Loeb 2: 256–259); see discussion by John Scarborough, "Drugs and Drug Lore in the Time of Theophrastus: Folklore, Magic, Botany, Philosophy and the Rootcutters," *Acta Classica* 69 (2006): 1–29, esp. 17–18.
44. Van Arsdall, unpubl. paper cited earlier.
45. Ginzberg, p. 366.
46. Ibid., p. 368.
47. From other sources we deduce that his sensitiveness led him to avoid saying menstrual blood.
48. Josephus, *Jewish War,* 7. 3. 180–185 (H. St. J. Thackeray, trans. in Loeb. 3: 556–559); and new translation by Gaalya Cornfeld (1982), p. 474.
49. Ibid., 7.3. 183.
50. For example, Rosemary A. Cotes, *Bible Flowers* (London: Methuen, 1904), p. 64, and Jeanne Rose, *Herbs and Things* (New York: Grossett and Dunlap, 1973), p. 81.
51. Josephus, *Jewish Antiquities,* 1. 8. 307 (Thackeray, 1: 148–149); and new translation by William Whiston (1999), p. 75.
52. The assertion that Baaras is mandrake is found in Christian Rätsch, "Einleitung," to: Alfred Schlosser, *Die Sabe vom Galgenmännlein im Volksglauben und in der Literatur* (Berlin: Express Edition, 1987), p. xviii.
53. R. Campbell Thompson, "The Migration of Assyrian Plant-names into the West," *The Classical Review* 38 (1924): 148.
54. Hildegard of Bingen, *The Book of Blessed Hildegard begins, Physica.* Bruce W. Hozeski, trans. (Boston: Beacon Press, 2002), 56, p. 51. See also Marvin H. Pope, *Song of Songs: A New Translation with Introduction and Commentary.* Anchor Bible Series (Garden City, NY: Doubleday, 1977), p. 649.
55. Plaut, Bamberger, and Hallo, p. 207.
56. Robert D. Biggs, *ŠA`.ZI.GA. Ancient Mesopotamian Potency Incantations* (=*Texts from Cuneiform Sources*, vol. 2, Locust Valley, NY: J. J. Augustin, 1967), p. 70.
57. http://www.sumerian.org/sumg-k.htm s.v>. (accessed 6/27/08). R. Campbell Thompson (1949, p. 255) believes *hašhur* is likely oak galls.
58. Moldenke and Moldenke, p. 139.
59. R. Campbell Thompson, "Assyrian Medical Texts," *Proceeding of the Royal Society of Medicine* 19 (1926): 56; on p. 117, Thompson has an enema with mandrake root but the affliction is indecipherable. On opium in antiquity, see John Scarborough, "The Opium Poppy in Hellenistic and Roman Medicine," in *Drugs and Narcotics*

in History. Roy Porter and Mikuláš, eds. (Cambridge: Cambridge University Press, 1995), pp. 4–23.

60. Markham J. Geller, *Renal and Rectal Disease Texts.* Vol. 7 of *MTU* (Berlin: de Gruyter, 2005), pp. 32–33, 44–45, 164–165, 176–177, 192–193, 198–199, 258–259.
61. Thompson, p. 58.
62. Ibid., p. 75.
63. Ibid., p. 77.
64. Thompson (1924; 1956 repr.), pp. 67, 187–189.
65. Ibid., p. 188.
66. *Larousse Encyclopedia of Mythology* (New York: Barnes and Noble, 1994), p. 163.
67. Robert S. Holtzman, "The Legacy of Atropos, the Fate Who Cut the Thread of Life," *Anesthesiology* 89/1(1998): 241–249; also see *Therapeutic Drugs.* 2nd ed., 2 vols. Colin Dollery, ed. (Edinburgh: Churchill Livingston, 1999), 1: A240–245 and Internet Drug List: http://www.rxlist.com/cgi/generic/atrop_cp.htm (accessed 7/3/08).
68. Wink, p. 29.
69. PDRHealth. Physicians' Desk Reference online: http://www.pdrhealth.com/drugs/rx/rx-mono.aspx?contentFileName=Lev1223.html&contentName=Levsin&contentId=302 (accessed 8/1/08).
70. AccuHist, see http://www.accuhist.com/ Yahoo.Health. http://health.yahoo.com/flu-medications/atropine-chlorpheniramine-hyoscyamine-phenylephrine-scopolamine/healthwise—d04872a1.html (accessed 8/1/08).
71. James A. Duke, *CRC Handbook of Medicinal Herbs* (Boca Raton, FL: CRC Press, 1985), p. 292.
72. Holtzman, 241.
73. *Therapeutic Drugs.* 2nd ed., in 2 vols. Colin Dollery, ed. (Edinburgh: Churchill Livingston, 1999), 1: A240–245.
74. Holtzman, 241.
75. John Scarborough, "Mandrake in Ancient Surgery," unpublished paper presented at the Society for Ancient Medicine jointly with the American Philological Association, January 7, 2006, Montreal, Quebec.
76. Monika Adt, Peter Schmuker, and Iris Müller, "The Role of Atropine in Antiquity and in Anaesthesia," in *The History of Anaesthesia.* Richard S. Atkinson and Thomas B. Boulton, eds. (London: Parthenon, 1987), p. 44.
77. *Diseases.* 2. 43. Paul Potter, trans. in *Hippocrates.* Loeb Classical Library (Cambridge, MA: Harvard University Press,1988) 5: 270–272 ; Littré, ed 7:66.
78. Robert Sallares, *Malaria and Rome: A History of Malaria in Ancient Italy* (Oxford: Oxford University Press, 2002), pp. 9–11.
79. Émile Littré, the great editor of the Hippocratic works, translated the term to mean "melancholia" or our mental depression, but I cannot see that precision in the Greek term.
80. *De locis in homine.* 39 (E. Littré ed., 6: 328–329).
81. *De mulierum affectibus.* 2. 199 (Littré, 8: 382).
82. *De fistulis.* 9 (Littré 6: 458).
83. *De locis in homine.* 39 (Littré, 6: 329).
84. Ibid. 1. 74 (L. 8: 163–164), which is followed by another pessary that "gives an even greater response" and contains a number of plant and animal drugs beginning with mandrake root. In the same treatise mandrake seed (*kokkos*') is infused in the uterus if the woman is in an anxious state (*cholōdēs*) (1. 80; L. 8:202).
85. Jackson, Bettty P., and Michael I. Berry, "Hydrooxytropane Tiglates in the Roots of *Mandragora* Species," *Phytochemistry* 12 (1973): 12/5 (1973): 1165–1166.

86. Dioscorides, *De materia medica,* 4. 75. 5 (Lily Y. Beck trans., p. 281).
87. Dioscorides, *De materia medica,* 4. 75.1 (Beck trans., p. 280).
88. Ibid., 2.100. 3: 52 and 2:75.
89. Ibid., 2. 75 (Beck trans., p. 115).
90. Peter V. Taberner, *Aphrodisiacs: The Science and the Myth* (Philadelphia: University of Pennsylvania Press, 1985), pp. 111–120 (on mandrake).
91. Michael McCormick, "Towards a Molecular History of Justinianic Plague," in *Plague and the End of Antiquity: The Pandemic of 541–750.* Lester E. Little, ed. (Cambridge: Cambridge University Press, 2007), p. 295.
92. *De morborum causis.* 3 (7:14 Kühn ed.); *De actutorum morborum victu.* 1. 32 (15:489 K.); *De simplicium medicamentorum temp. ac fac.* 1. 13 (11:404 K.), 1. 22 (11:421 K.); 5. 14 (11: 751 K.); 7. 12. 4 (11: 766–767).
93. *De simpl. med. temp ac fac.* 2. 18 (11: 596 K).
94. Ibid., 5. 19 (11: 766–767).
95. *De morborum causis.* 3 (7: 14 K).
96. Soranus, *Gynecology,* Oswei Temkin, trans. (Baltimore: Johns Hopkins University Press, 1956), 1. 35.
97. Ibid.
98. *Historia naturalis.* Bk. 26: 12. 24, 60, 93, 66, 104–105, 74. 121, 87. 145.
99. Ibid. 26: 91.162.
100. Ibid. 26: 91. 162 (W. H. S. Jones, trans. in Loeb).
101. Ibid. 14: 21. 116 (Jones trans.)
102. *De medicina.* 3. 18. 12.
103. Columella. 10. 20.
104. Xenophon, *Banquet,* 2. 24 (O. J. Todd, trans., 1992 repr., p. 555.).
105. Apuleius, *Metamorphoses,* 10. 11 (J. Arthur Hanson, trans. in Loeb Classical Library, 2 vols. [Cambridge, MA: Harvard University Press, 1989], 2: 237).
106. References with quotations in English given by Randolph, 485–537, esp. 509–512.
107. Polyaenus, *Stratagems,* 8. 23. 1 (Peter Krentz and Everett Wheeler, eds. 2: 754).
108. Aristotle, *On Sleep and Dreams,* 12. 456b28; Isidore, *Origines,* 17. 9. 30.
109. Frontinus, *Stratag,* 2.5.12 (Charles Bennett, trans. in Loeb Classical Library), pp. 138–139.
110. W. Geoffrey Arnott, *Alexis. The Fragments. A Commentary* (Cambridge Classical Texts and Commentaries, vol. 31; Press Syndicate of the University of Cambridge Press, 1996), pp. 419–420.
111. Caelius Aurelianus, *On Chronic Diseases,* 1. 147 (I. E. Drabkin trans., p. 537).
112. Plutarch, *Moralia: Quomodo adolescens poetas audire debeat,* 15 F (Frank Cole Babbit, trans. in Loeb Classical Library, 1: 80–81).
113. Hesychius *s.v.* 3: 69.
114. Tacitus, *Annals,* 12. 67 (see notes by John Jackson in Loeb ed. of Tacitus 3: 414–415); Suetonius, *Claudius,* 44. 1–3; Dio. 56. 30. 1–3.
115. John Timbrell, *The Poison Paradox* (Oxford: Oxford University Press, 2005), p. 148; John Mann, *Murder, Magic, and Medicine* (Oxford: Oxford University Press, 1994), p. 24; Wink, p. 20; one needs only to Google mandrake, poisons, Livia, and Agrippina to see the number of references that pass these "facts" as reality.
116. Charlesworth, "Livia and Tanaquil," *The Classical Review* 41/2 (1927): 55.
117. Van Arsdall, p. 15 (in manuscript).
118. Gohar Muradyan, *Physiologus: The Greek and Armenian Versions with a Study of Translation Technique* (Leuven: Peeters, 2005).

119. *The Bestiary. A Book of Beasts Being from a Translation from a Latin Bestiary of the Twelfth Century,* T. H. White, ed. (New York: Putnam's Sons, 1960 ed.), p. 27; see also an early English version, *The Middle English Physiologus.* Hanneke Wirtjes, ed. (Oxford: Early English Text Society, 1991), lines 493–512, p. 18. The origins of a connection between the elephant and Adam and Eve remain a mystery to me. Ambrose (Hexameron 5 Savage trans., pp. 246–249) either borrowed from Physiologus or the two borrowed from the same sources for the elephant but Ambrose did not include Adam, Eve, and the Garden of Eden. See discussion of sources by Michael J. Curley, *Physiologus* (Austin: University of Texas Press, 1979).
120. Van Arsdall, pp. 57–74.
121. *Theobaldi Physiologus,* P. T. Eden, trans. and ed. (Leiden: Brill, 1972), p. 67, line 24; Latin text is available on Corpus Scriptorum Latinorum. Theobaldus *Physiologus* on Web site: http://www.forumromanum.org/literature/theobaldus/physiologus.html (accessed 7/30/08).
122. Hugo Rahner, *Greek Myths and Christian Mystery.* Brian Battershaw, trans. (New York: Harper and Row, 1963; German ed., 1957), p. 257.
123. Francis J. Carmody, "Physiologus Latinus Version Y," *University of California Publications in Classical Philology* 12 (1944): 118.
124. Hildegard, *Physica,* 1. 56 as cited by William Scott Shelley, *The Elixir: An Alchemical Study of the Ergot Mushrooms* (Notre Dame: Crosse Cultural Publications, 1995), p. 129.
125. Max Wellmann, "Allgemeiner Charakter des Physiologos. Zeit und Ort seiner Entstehund," in *Philologus. Zeitschrift für das klassische Altertum,* vol.22 (Leipzig: Dieterich'sche Verlagsbuchhandlung, 1930), 22/1: 41–42.
126. Ibid. p. 41 [citing Ibn al-Baithar (ed. Sontheimer, II, 14, 606).
127. Rahner, p. 257.
128. Ibid., p. 256.
129. I am grateful to Anne Van Arsdall for observing the elephant's significance in Christian lore.
130. Debra Hassig, *Medieval Bestiaries: Text, Image, Ideology* (Cambridge: Cambridge University Press, 1995), pp. 129–144.
131. T. H. White, translation, *The Bestiary: A Book of Beasts* (New York: G. P. Putnam's Sons, 1954), p. 27; Theobaldus wrote another bestiary version in the eleventh century in which he probably collapsed the details but said that Adam "fell because of the wood of a tree (*per lignum*)," not naming the tree but tying the disobedience to the tree of knowledge. See, *Theobaldi Physiologus,* P. T. Eden, trans. and ed. (Leiden: Brill, 1972), p. 67.
132. Also as expressed in an English version found in Oxford Bodl. MS 764: "When the woman ate of the tree, that is, gave the herb mandragora which brought understanding to her husband, she became pregnant and for that reason left paradise. For as long as they were in paradise, Adam did not know her in the flesh." In *Bestiary: Being from an English Version of the Bodleian Library*... (Woodbridge, UK: Boydell and Brewer, 1992), pp. 40–41.
133. E. A. Wallis Budge (translation and commentary), *The Book of Medicines: Ancient Syrian Anatomy, Pathology, and Therapeutics* (London: Kegan Paul, 2002 repr.), p. 713; mandrake as a medicine is discussed in prescriptions on p. 274 (diseases of the chest), 370 (ulcers of stomach and various abdominal disorders), 371 (relaxing the stomach, severe nausea, diarrhea, and liver disease—said to be very powerful).
134. Samuel S. Kottek, *Medicine and Hygiene in the Works of Flavius Josephus* (Leiden: Brill, 1994), p. 132.

135. Preuss, p. 463.
136. Augustine, *Contra Faustum manichaeum*. 22. 56 (in *Patrologiae cursus completus. Series Latina,* 42: 435).
137. Lumír O. Hanuš, Tomáš Řezanka, Jaroslav Spížek, and Valery M. Dembitsky, "Substances Isolated from *Mandragora* Species," *Phytochemistry* 66 (2005): 2415–2416.
138. Colette Bouchez (reviewed by Brunilda Nazario), WebMD: "Infertility and Reproduction Health Center": http://www.webmd.com/infertility-and-reproduction/features/infertility-stress (accessed 7/24/08).
139. A. D. Doman, P. C. Zuttermeister, and R. Friedman, "The Psychological Impact of Infertility: A Comparison with Patients with Other Medical Conditions," *Journal of Psychosomatic Obstetrics and Gynaecology* 14 (1993 special issue): 45–52; see also, Markus S. Kupka et al., "Stress Relief after Infertility Treatment-spontaneous Conception, Adoption, and Psychological Counseling," *European Journal of Obstetrics and Gynecology Reproductive Biology* 110/2 (2003): 190–195.
140. Annett L. Stanton, Sharon Sears, Marci Lobel, and Robyn Stein DeLuca, "Psychosocial Aspects of Selected Issues in Women's Reproductive Health: Current Status and Future Directions," *Journal of Consulting and Clinical Psychology* 70/3 (2002): 758 (article has numerous references).
141. Frederick J. Kakis, *Drugs: Facts and Fiction* (New York: Franklin Watts, 1982), p. 207.
142. Bouchez.
143. Robyn R. Wilborn and Brian K. Whitlock; Diplomates of the American College of Theriogenologists (veterinarians dedicated to animal reproduction); Dept. of Clinical Sciences, Auburn University College of Veterinary Medicine, Auburn, AL. I am grateful to Drs. Wilborn and Whitlock for helping me understand the effects of stress on ovulation and to Dr. Carlos Pinto of the School of Veterinary Medicine at North Carolina State University for his assistance in contacting fertility experts.
144. S. N. Kalantaridou, et al., "Stress and the Female Reproductive System," *Journal of Reproductive Immunology* 62 (2004): 61.
145. Wolfgang Schmidbauer, "Die magische Mandragora," *Antalos* 10 (1969): 284, suggested that the soporific quality of mandrake may be a natural explanation of the fertility-enhancing quality rather than the "magic" explanation.
146. Contrary to Harrison, p. 91, who believed that Rachel's conception was attributable to "emotional satisfaction and psychological suggestion."
147. Machiavelli, *Mandragola,* David Sices and James B. Atkinson, eds. and trans. (Hanover and London: University Press of New England, 1985), pp. 158–159.
148. D. C. Allen, "Donne on the Mandrake," *Modern Language Notes* 74/4 (1959): 396. I am grateful to John Hester for giving me this reference.
149. *English Poetry (1170–1892).* John Matthews Manley, ed. (Boston: Ginn, 1907), p. 153.

CHAPTER 4

1. Hubert M. Martin, Jr., "Artemis," in *ABD,* 1: 464–465; see also Wernicke, "Artemis," in *Paulys Real-encyclopädie der classischen Altertumswissenschaft* (Stuttgart: Metzler, 1903–) 2/1: 1335–1440.
2. Philip Smith, "Mausoleum," in *Dictionary of Greek and Roman Antiquities.* William Smith, ed. (London: Murray, 1875), pp. 744–745, with citations to classical sources.

3. *Gerard's Herball: The Essence thereof distilled by Marchus Woodward. From edition of Th. Johnson, 1636* (New York: Crescent Books, 1985 ed.), p. 254.
4. Manuela Dunn Mascetti, *Artemis: Goddess of the Hunt and Moon* (San Francisco: Chronicle Books, 1996), pp. 16–18.
5. Heinrich von Staden, "Spiderwoman and the Chaste Tree: The Semantics of Matter," *Configurations* 1/1 (1993): 46–47.
6. Paul Cartledge, *Sparta and Lakonia: A Regional History, 1300–362 BC* (London: Routledge, 2002), pp. 308–309; P. Cartledge, *Spartan Reflections* (Berkeley and Los Angeles: University of California Press, 2001) p. 172. Pausanias, Plutarch, Xenophone, and Plato all attested to the cruel ritual.
7. Claude Calame, Derek Collins, and Janice Orion, *Choruses of Young Women in Ancient Greece: Their Morphology, Religious Role, and Social Functions* (Lantham, MD: Rowman & Littlefield, 2001), pp. 98–99; Angelo Brelich, *Paides e Parthenoi* (Rome: Edizioni dell'Ateneo, 1969), p. 240, who said that they learned "attività fondamentali della donna."
8. *Dictionary of World Myth.* Peter Bently, ed. (New York: Facts on File, 1995), p. 25.
9. Plants. United States Department of Agriculture. http://plants.usda.gov/java/invasiveOne?startChar=A (accessed 8/25/08).
10. Colin W. Wright, *Artemisia* (London and New York: Taylor and Francis, 2002), p. ix; Hongwe Yu and Shouming Zhong, "*Artemisia* Species in Traditional Chinese Medicine and the Discovery of Artemisinin," in *Artemisia*, p. 156.
11. Xenophone, *Anabasis*, 1. 5. 1 (Carleton Brownson, trans. in Loeb, 1:286).
12. Michael Zohary, *Plants of the Bible* (Cambridge: Cambridge University Press, 1982), p. 184.
13. Ahmed A. Mahmoud and Ahmed A. Ahmed, "α-Pinene-Type Monoterpenes and Other Constituents from *Artemisia suksdorfii*," *Phytochemistry* 67/19 (October, 2006): 2103–2109, esp. 2104; Samir A. M. Abdelgaleil et al., "Bioactivity of Two Major Constituents Isolated from the Essential Oil of *Artemisia judaica* L.," *Biosource Technology* 99/12 (2007): 5947–5950; James Duke, *Duke's Handbook of Medicinal Plants of the Bible* (Boca Raton, FL: CRC Press, 2008), pp. 50–52.
14. Caner, Ayşe et al., "Comparison of the Effects of *Artemisia vulgaris* and *Artemisia absinthium* growing in Western Anatolia against Trichinellosis (*Trichinella spiralis*) in Rats," *Experimental Paraistology* 1119/1 (2008): 173–179.
15. Moldenke, pp. 48–50.
16. Thompson, *DAB*, pp. 233, 359–362; JoAnn Scurlock and Burton R. Andersen, trans. and commentary, *Diagnoses in Assyrian and Babylonian Medical Analyses* (Urbana: University of Illinois Press, 2005), pp. 81–87, offer a number of parasitical infections but there is little in the way of therapy presented.
17. Ebers Papryus, in *Papyros Ebers. Das älteste Buch über Heilkunde.* H. Joachim, ed. and trans. (Berlin: Walter de Gruyter, 1973), pp. 12–13, 17.
18. Lise Manniche, *An Ancient Egyptian Herbal* (Austin: University of Texas Press, 1989), p. 80; Ebers in Joachim ed., pp. 29, 33, whose recipes are slightly different and include other afflictions, such as abdominal pain, and nontranslated problems.
19. Ebers in Joachim, pp. 3, 5–6, 9, 18–20, 24, 26, 29, 37, 41–44, 61, 126, 128–129, 133, 142, and 149.
20. I cite a number of studies published in one volume Wright, pp. 9, 54–55, 66, 68–69, 80, 109, 110–111, 114, 131, and 145.
21. Cheryl Lans, Nancy Turner, Yonya Khan, and Gerhard Brauer, "Ethnoveterinary Medicines Used to Treat Endoparasites and Stomach Problems in Pigs and Pets in British Columbia," *Veterinary Parasitology*, 148/1–2 (September 2007): 325–340.

22. R. X. Tan, W. F. Zheng, and H. Q. Tang, "Biologically Active Substances from the Genus *Artemisia,*" *Planta Medica* 64 (1998): 295–302; James A. Duke, *CRC Handbook of Medicinal Herbs* (Boca Raton, FL: CRC Press, 1985), pp. 66–70; Wright, in entirety.
23. H. J. Woerdenbag and N. Pras, "Analysis and Quality Control of Commercial *Artemisia* Species," in *Artemisia,* p. 54.
24. Ibid.
25. Dioscorides, *De materia medica,* 3. 23 (Lily Y. Beck, trans., p. 189).
26. Cyril P. Bryan, trans. *Ancient Egyptian Medicine: The Papyrus Ebers* (Chicago: Ares, 1974), p. 40; Bryan translates other prescriptions using wormwood as a laxative and for diseased toes (pp. 45, 62).
27. Pliny. *NH.* 25. 36. 73.
28. Pseudo-Apuleius, *Herbarius,* 12. 2 (Ernest Howald and Henry Sigerist, eds., p. 45).
29. Marcellus, *De medicamentis,* 20. 83 (Niedermann ed., 1: 348).
30. Dioscorides, *De materia medica,* 3. 113 (Beck, trans., p. 233).
31. For example, http://dreampharm.com/garlic/mugwort.asp and http://www.pmsice.com/Ingredients.cfm (accessed 9/8/08).
32. Dreampharm. http://dreampharm.com/garlic/mugwort.asp (accessed 9/8/08).
33. PMS Ice. http://www.pmsice.com/Ingredients.cfm (accessed 9/8/08).
34. Sarí Harrar and Sara Altshul O'Donnell, *The Women's Book of Healing Herbs: Healing Teas, Tonics, Supplements, and Formulas* (Emmaus, PA: Rodale Press, 1999). Motherwort (or Mother's Herb) is identified as lion's tail (*Leonurus cardiaca* L.). Tony Hunt, *Plant Names of Medieval England* (Cambridge: D. S. Brewer, 1989), p. 295, identifies some twelve species passing under the name of mother herb in English during the medieval period. Generally, however, in the Middle Ages the "mother herb" was also known a mugwort (*Artemisia vulgaris* L.).
35. John Riddle, *Eve's Herbs: A History of Contraception and Abortion in the West* (Cambridge, MA: Harvard University Press, 1997), pp. 47–48, 56, 83, 86, 89–92, 98, 103–104, 114, 122–124, 137, 154, 161.
36. EDrugDigest: http://www.drugdigest.org/DD/DVH/HerbsWho/0,3923,4064|Wormwood,00.html (accessed 10/5/08).
37. Soranus, *Gynecology,* Oswei Temkin, trans. (Baltimore: Johns Hopkins University Press, 1956), 1. 64 (pp. 66–67).
38. Ibid., 1. 64–65 (pp. 66–67).
39. Ibid., 1. 52 (p. 53).
40. Ibid., 4. 13; Hippocrates, *On Women's Diseases,* 1. 77 (Littré ed., vol. 8: 170–172).
41. Hippocrates, *On Women's Diseases,* 2. 135 (Littré, 8:306).
42. Hippocrates. *On Female Sterility,* 3. 230 (Littré, 8: 443).
43. Hipocrates, *Regimen in Acute Diseases,* 34 (Potter ed., 6: 298); *Diseases* 3. 11 (Potter 6: 24); *Internal Afflictions,* 52 (Potter 6: 250).
44. Galen. *De simpl. med. temp. ac fac,* 6. 1 (Kühn ed., 11: 798–807, 839–840; *De methodo medendi,* 11. 16 (expel poisons; K. 10: 789–790); 8. 5 (K. 10: 572); *De sanitate tuenda,* 6. 7 (K. 6: 428); *De alimentorum facultatibus.* 3. 32 (K. 6: 731); *De antidotis.* 2.7 (K. 14: 140).
45. Galen, *De simpl. med. temp. ac fac,* 6.1 (11: 804).
46. Celsus, *De medicina,* for *abrotonum* (southernwood) 4. 8.3–4 (1: 386 Spencer ed.); 5. 3. 5, 9. 11 (2: 6, 10); absinthe: 2. 24. 2 (2: 30); 2. 24. 2 (1: 204); 2. 31. 1 (1: 210); 3. 21. 6 (1: 316); 3. 24. 2 (1: 340); 4. 7. 3 (1: 386); 4. 12. 2 (1: 398); 4. 12. 4 (1: 398); 4. 12. 6 (1: 401); 4. 15. 3 (1: 414); 4. 18. 8 (1: 422; 5. 25. 16 (2: 64); 4. 24. 2 (1: 438); 4. 15. 2 (1: 412).

47. Ibid., 3. 21. 7 (1:316).
48. Pliny, *Natural History,* 21. 92. 160; 22. 30. 65; 23. 16. 52; 23. 57. 123; 23. 76. 124, 146; 26. 58.91; 26. 90. 151. 26. 90. 159; 28. 58. 203; 29. 39. 133.
49. Ibid., 14.19. 109; 19. 58. 179; 20. 7. 15; 21. 79. 135.
50. Ibid., 25. 34. 73.
51. Scribonius Largus, *Compositiones,* 78, 188, 192, 228 (George Helmreich, ed. Leipzig: Teubner, 1887, pp. 33, 77, 78, 91); Caelius Aurelianus, *On Chronic Diseases,* 4. 122, 124; on southernwood, Nicander. *Theriaca* 66 (for aroma therapy); 92; 574; *Alexipharmaca,* 46 (antidote to gypsum toxin); on absinthe, *Alexipharmaca,* 298 (as anodyne).
52. Macer, *Herbarum* [Paris]: Bacquelier, [*ca.* 1515], no foliation.
53. Al-Kindī. *Formulary* (Martin Levy, trans.) p. 296.
54. Quoted by Martin Levy, *Al-Kindī,* p. 296.
55. Avicenna, *Liber Canonis,* Bk. 2, Tract. 2., cap. 2 (Venice, 1507), fol. 88r&v.
56. S. D. Kamat, *Studies on Medicinal Plants and Drugs in Dhanvantari-Nighantu* (Delhi: Chaukhamba Sanskrit Pratishthan, 2002), pp. 171–172.
57. *Bhela-Samitā. Text with English Translation, Commentary, and Critical Notes.* K. H. Krishnamurthy, trans; Priya Vrat Sharma, ed. (Varanasi, India: Chaukhambha Visvagharai, 2000); *Sushruta Samhita.* Kaviraj Kunjalal Bhishabratna, ed. and trans. in 3 vols. (Varanasi: Chowkhamba Sanskriti Series, 1963).
58. Quoted through Rufinus, *Herbal.* fol. 17 (Lynn Thorndike ed., p. 2).
59. Ibid.
60. Hildegard, *Physica,* 109. in *Patrologiae.* vol. 197, cols. 1172b–1173b.
61. Platearius, *Circa instans,* Hans Wölfel ed. (Berlin: A. Prelipper, 1939), pp. 14–15.
62. Schneider, vol. 5/1, p. 134.
63. Nicholas Culpeper, *Pharmacopoeia Londinensis or, The London Dispensatory* (London: 1720), p. 17.
64. Ibid., 5/1, p. 135.
65. Ibid., pp. 135–136.
66. Jacob Grimm, *Teutonic Mythology.* James Stallybrass, trans. in 4 vols. (Gloucester, MA: Peter Smith, 1976), 3: 1211.
67. Ibid., 3: 1211–1212.
68. Ibid., 4: 1679.
69. Ibid.
70. Ibid.
71. Francis B. Brévart, "'Mother of All Herbs': The Magical Plant Mugwort (*Artemisia vulgaris* L.) in Medieval German Wonder Drug Literature," in *Er ist ein wol gevri-under man. Essays in Honor of Ernst S. Dick.* (Hiledesheim: Olms, 2009), pp. 44. 43–72.
72. *New York Times*, February 1, 2008, A4.
73. BBC News, February 8, 2003: http//newsvote.bbc.co.uk/mpapps/pagetools/print/news. bbc.co.uk/2.html (accessed 8/25/08).
74. Robert Sallares, *Malaria and Rome: A History of Malaria in Ancient Italy* (Oxford: Oxford University Press, 2002), pp. 9–10; Mirko D. Grmek, *Diseases in the Ancient World.* Mireille Muellner and Leonard Muellner, trans. (Baltimore: Johns Hopkins University Press, 1989), pp. 275–283.
75. *Management of Severe Malaria: A Practical Handbook* (Geneva: World Health Organization, 2000).
76. Microbiologybytes Web site: http://www.microbiologybytes.com/introduction/Malaria.html (accessed 8/26/08).

77. Celsus, *De medicina,* 3. 3. 2; Hippocrates, *Epidemics,* 1. 24. 10 (W. H. S. Jones, trans. in Loeb, *Hippocrates,* 1:183).
78. Celsus, *De medicina,* 3. 3. 2 (W. G. Spencer, trans. in Loeb 1: 227).
79. On fevers in antiquity (without attempting to relate to modern medicine), see Wesley D. Smith, "Fever Pathology in *Epidemics* 5 and 7," in W. F. Bynum and Vivian Nutton, eds., *Theories of Fever from Antiquity to the Enlightenment* (London: Wellcome Institute, 1981), pp. 1–18.
80. Diogenes Laertius, 8. 70; see PE II A. 2. (1923) col. 1281.
81. Hippocrates, *"Airs, Water, Places,"* 7. 9–10 (Jones trans.).
82. I. A. McGregor, "Malaria," in *The Wellcome Trust Illustrated History of Tropical Diseases.* F. E. G. Cox, ed. (London: Wellcome Trust, 1996), p. 232.
83. Ibid., p. 232; Colin W. Wright, ed. *Artemisia.* (London and New York: Taylor and Francis, 2002), p. ix; Phyllis Williams Lehmann, *Statues on Coins of Southern Italy and Sicily in the Classical Period* (New York: H. Bittner, 1946), pp. 15–16. Lehmann gives a bibliography of previous discussants about the interpretation of the coin's meaning. See also, Stanley Lane-Poole, *Coins and Medals: Their Place in History and Art* (London: Elliott Stuck, 1885), pp. 25–26.
84. *The Yellow Emperor's Classic of Medicine.* Maoshing Ni, trans. (Boston and London: Shambhala, 1995), p. 138.
85. Ibid. p. 141.
86. Liu An, *Divine Farmer's Materia medica: A Translation of the Shen Nong Ben Cao Jing.* Yang Shouzhong, ed. (Boulder, CO: Blue Poppy Press, 1998), p. 29. Elsewhere in the same work (p. 33) is the herb *bai hao* identified as *Artemisia argyi* Levi. whose medicinal qualities are much the same as those of the other species.
87. "Translation of MSI.E," by Donald J. Harper, *Early Chinese Medical Literature: The Mawangdui Medical Manuscripts* (London and New York: Kegan Paul, 1998), p. 272.
88. Elizabeth Hsu, "Reflections of the 'Discovery' of the Antimalarial *Qinghao,*" *British Journal of Clinical Pharmacology* 61/6 (2006): 667.
89. William R. Burns, "East Meets West: How China Almost Cured Malaria," *Endeavor* (prepublication, 2008; accessed via Science Digest, *www.sciencedigest.com* 8/31/08).
90. Title in translation: "Antimalarial Coordinating Research Group: Antimalarial Studies on Qinghaosu," *Chinese Medical Journal* 92 (1979): 811–816. A few pages of the 1979 article are reproduced on this Internet site called Panyu Tiger. http://panyutiger.blogspot.com/2008/05/chinese-medical-journal-paper-1979.html (accessed 8/31/08).
91. Hsu, pp. 668–669.
92. A. G. Namdeo, K. R. Mahadik, and S. S. Kadam, "Antimalarial Drug—*Artemisia annua,*" *Pharmacognosy Magazine* 2/6 (2006): 106.
93. S. R. Meshnick, T. E. Taylor, and S. Kamchonwongpaisan, "Artemisinin and Antimalarial Endoperoxides: From Herbal Remedy to Targeted Chemotherapy," *Microbiological Reviews* 60/2 (1996): 303.
94. Hongwen Y and Shouming Zhong, "*Artemisia* Species in Traditional Chinese Medicine and the Discovery of Artemisinin," in *Artemisia,* p. 150.
95. Dae-Kyun Ro et al., "Production of Antimalarial Drug Precursor Artemisinic Acid in Engineered Yeast," *Nature* 440 (April 13, 2006): 940–943.
96. Karin Räth, Katja Taxis, Gitta Walz, Christoph H. Gleiter, Shu-Ming Li, and Lutz Heidi, "Pharmacokinetic Study of Artemisinin after Oral Intake of a Traditional Preparation of *Artemisia annua* L. (Annual Wormwood)," *American Journal of Tropical Medicine and Hygiene* 70/2 (2004): 128–132.

97. Ibid.
98. Hsu, p. 669.
99. Wright, p. ix; Yu and Zhong, "*Artemisia* Species," p. 156.
100. Daniel L. Klayman, "*Qinghaosu* (Artemisinin): An Antimalarial Drug from China," *Science* n.s. 2228/4703 (1985): 1049.
101. Namdeo, Mahadik, and Kadam, p. 108.
102. Carola Kraft et al., "*In Vitro* Antiplasmodial Evaluation of Medicinal Plants from Zimbabwe," *Phytotherapy Research* 17 (2003): 123–128 quoting from p. 128.
103. B. Cubukcu et al., "*In Vitro* Antimalarial Activity of Crude Extracts and Compounds from *Artemisia abrotanum* L.," *Phytotherapy Research* 4 (1990): 203–204.
104. Neena Valecha. S. Biswas, V. Badoni, K. S. Bhandari, O. P. Sati, "Antimalarial Activity of *Artemisia Japonica, Artemisia Maritima, and Artemisia Nilegaricia*," *Indian Journal of Pharmacology* 26 (1994): 144–146.
105. Merlin Willcox et al., "*Artemisia annua* as a Traditional Herbal Antimalarial," in *Traditional Medicinal Plants and Malaria*, eds. Merlin Willcox, Gerard Bodeker, and Philippe Rasanaro (Boca Raton, FL: CRC Press, 2004), p. 50.
106. *Diseases* 2. 40 (Potter trans., 5. 255–257).
107. Sallares, pp. 18–19.
108. For a recent review of older views, see Sallares, pp. 1–5; Cheston B. Cunha and Burke A Cunha, "Brief History of the Clinical Diagnosis of Malaria: From Hippocrates to Osler," *Journal of Vector Borne Disease* 45 (2008): 194–199.
109. Wolfgang Schneider, *Lexikon zur Arzneimittelgeschichte,* 7 vols. in 9 pts. (Frankfurt a. M.: Govi-Verlag, 1974), 5/1: 295–304.
110. *Syriac Book of Medicines,* p. 337, Budge trans.
111. *Hortulus,* 9 (Raef Payne, trans., p. 41).
112. Al-Kindī, *The Medical Formulary.* Martin Levy, trans. (Madison: University of Wisconsin Press, 1966), p. 233.
113. Avicenna, *Liber Canonis* Bk. 2. Tract. 2., cap. 2 (fol. 88v, 1507 ed.).
114. Constantine, *De gradibus.* Munich Bay. Staatsbibliothek MS lat. 267, fol. 119.
115. Various modern writers cite Paracelsus' use of artemisia oil to treat malaria but I have been unable to trace the citation to a work of Paracelsus; see, for example, Paul Herrewijn, Adrianna M. van Oosten, and Paul G. M. Piron, *Natural Terpenoids as Messengers* (Dordrecht: Kluwer Academic, 2001), p. 174; Doris Lanier, *Absinthe: The Cocaine of the Nineteenth Century* (Jefferson, NC: McFarland, 1995), p. 3.
116. See R. E. Latham, *Revised Medieval Latin Word-List* (London: British Academy, 1965), *s. v.*
117. Macer, *Herbarium.* s.v *abrotinum.*
118. Nicholas Culpeper, *Complete Herbal* (London: W. Foulsham, n.d.), p. 394.
119. Ibid.
120. Dioscorides, *De materia medica,* 3. 23. 3 (Beck trans., p. 189).
121. John Riddle, "Ancient and Medieval Chemotherapy for Cancer," *Isis.* 76/3 (1985): 319–330.
122. *New York Times*, February 1, 2008, A4.
123. Sallares, p. 49 (with references).
124. Federico Borca, "Towns and Marshes in the Ancient World," in *Death and Disease in the Ancient City.* Valerie M. Hope and Eireann Marshall, eds. (London and New York: Routledge, 2000), pp. 74–83.
125. Vivian Nutton, "Medical Thoughts on Urban Pollution," in *Death and Disease in the Ancient City*, p. 67.
126. Dioscorides, *De materia medica,* 3. 23 (Beck trans., p. 189).

127. Ibid.
128. Tan, Zheng, and Tang, p. 300 (with other references); Stephen O. Duke, Rex N. Paul, and Lee S. Mark, "Terpenoids from the genus Artemisia as Potential Pesticides," in *Biologically Active Natural Products. Symposium. American Chemical Society* 194 (1988/3802): 318–334.
129. *Nature's Medicines That Heal* (Washington: National Geographic Society, 2000), p. 343.
130. Hélène Chiasson et al., "Acaricidal Properties of *Artemisia absinthium* and *Tanacetum vulgare* (Asteraceae) Essential Oils Obtained by Three Methods of Extraction," *Journal of Economic Entomology* 94/1 (2001): 167–171.
131. Herodotus, *Persian Wars,* 2. 94–95 (George Rawlinson trans., pp. 160–161).
132. John Scarborough, *Medical Terminologies: Classical Origins* (Norman: University of Oklahoma Press, 1992), p. 111; Ian C. Beavis, *Insects and Other Invertebrates in Classical Antiquity* (Exeter, UK: University of Exeter Press, 1988), pp. 228–231.
133. Horace, *Epodes,* 9. 15–16, Bennett ed. Beavis (p. 234) says that mosquito nets were introduced to the Romans about the time of Augustus and cited this quotation by Horace. Merely because Horace has the first incidental surviving mention of the nets is not evidence of when they were introduced, especially given the paucity of information about daily life in the Republican era.
134. Paulus, Epigrams fragment, 9. 765 in *Greek Anthology* (W. R. Paton trans., 1: 412–413).
135. Ibid.
136. Propertius, *Elegies,* 3. 215.
137. Agathias Scholasticus, Epigrams fragment, 9. 766 in : *Greek Anthology* (W. R. Paton trans., 1: 412–413).
138. *Geoponica.* 13 (IΓ). 11. 1–10 (Henry Beckh, ed., pp. 396–387).
139. Grieve's *Modern Herbal,* pp. 858–859.
140. Thomas Tusser, *Five Hundred Points of Good Husbandry....* William Maver, ed. (London: Lackington, Allen, 1812), p. 172.
141. Marcus Terentius Varro, *Rerum rusticarum,* 1. 12. 2 (William D. Hooper and Harrison B. Ash, trans., p. 208).
142. Columela, *Rei rusticate,* 1. 5. 3 (Harrison B. Ash, trans., pp. 60–61).
143. Ibid., 1. 5. 6 (pp. 62–63).
144. Sallares, pp. 168–191.
145. Joseph Bosworth, *An Anglo-Saxon Dictionary* (London: Oxford University Press,1954), p. 1209 s.v.
146. "Gin" and "Cinchona" in *Encyclopedia Britannica On-line.* s. v.
147. Dioscorides, *De material medica,* 5. 6. 16 (Beck trans., p. 337).
148. Ibid., 3. 23 (Beck, p. 189).
149. *Nature's Medicine,* p. 343; also in Grieve, p. 858.
150. Ian S. Hornsey, *A History of Beer and Brewing* (Cambridge: Royal Society of Chemistry, 2003), pp. 417, 423–424.
151. Ibid., p. 424.
152. *Encyclopedia of Occultism and Parapsychology.* J. Gordon Melton, ed. (Detroit: Gale Research, 2001), 2: 1177.
153. Culpeper, p. 394.
154. Phil Baker, *The Dedalus Book of Absinthe* (Dedalus Books, 2001), pp. 104–105; Jad Adams, *Hideous Absinthe: A History of the Devil in a Bottle* (Madison: University of Wisconsin Press, 2004), pp. 20–21.
155. *Times,* April 26, 1872, p. 10, as cited by Adams, p. 20.

156. http://oxygenee.com/absinthe-effects.html (accessed 10/21/08).
157. Pierre Duplais, *A Treatise on the Manufacture and Distillation of Alcoholic Liquors.* M. McKennie trans. from French (Philadelphia, PA: Baird, 1871), p. 244.
158. Adams, p. 2.
159. Varro E. Tyler, *The Honest Herbal: A Sensible Guide to the Use of Herbs and Related Remedies.* 3rd ed. (New York: Pharmaceutical Products Press, 1993), p. 322; Duke, *CRC Handbook* (1983), pp. 66–67.
160. J. P. Meschler, and A. C. Howelett, "Thujone Exhibits Low Affinity for Cannabionoid Receptors but Fails to Evoke Cannabiminmetic Responses," *Pharmacology Biochemistry and Behavior* 62/3 (1999): 473–780.
161. J. del Castillo, M. Anderson, and G. M. Rubottom, "Marijuana, Absinthe and the Central Nervous System," *Nature* 253 (January 31, 1975): 365–366.
162. Duke (1983), p. 67.
163. *Dispensatory of the United States of America.* 25th. ed. (Philadelphia, PA: Lippincott, 1955), p. 1523.
164. *For Whom the Bell Tolls* (New York: Charles Scribner's, 1940), p. 50.
165. Dietrich Blumer, "The Illness of Vincent van Gogh," *American Journal of Psychiatry* 159/4 (2002): 519–526.
166. Tiln Rekand and Ilmer Sulg, "Absint og den kunstneriske kreativitet," *Tidsskrift for den Norske laegeforening tidsskrift for praktish medicinem ny reakke,* 123//1 (2003): 70–73.
167. Phillip E. M. Smith, "Absinthe Attacks," *Neurology and Art* 6 (2006): 376–381; W. N. Arnold, "Vincent van Gogh and the Thujone Connection," *Journal of the American Medical Association* 260/20 (1988): 3042–3044.
168. Geocities:http://www.geocities.com/heartland/bluffs/8336/robertservice/shooting.html (accessed 10/10/2008).
169. King James translation.
170. Moldenke, pp. 49–50.
171. Barnaby Conrad, *Absinthe: History in a Bottle* (San Francisco: Chronicle Books, 1988) pp. 1–2.
172. Adams, p. 1.
173. As quoted by Conrad, p. vii.

CHAPTER 5

1. Robert G. Hall, "Circumcision," in *ABD,* 1: 1025–1031.
2. J. F. Nunn, "The Origins of Anaesthesia," in *History of Anaesthesia.* Richard S. Atkinson and Thomas B. Boulton, eds. (London: Parthenon, 1987), pp. 21–22. Slightly different translation in John Wilson, "Egyptian Rituals and Incantations," in *ANET,* p. 326.
3. Nunn, p. 24; see also Ann Macy Roth, *Egyptian Phyles in the Old Kingdom: The Evoluion of a System of Social Organization* (Chicago: Oriental Institute of the University of Chicago, 1991), pp. 65–68, who raises the possibility that the knife's action may have been to cut the pubic hairs before circumcision.
4. Hall, *ABD,* 1: 1025.
5. The fusion thesis based on an alleged vision of world brotherhood was pronounced by William Tarn but critics to the sweeping, idealistic theory emerged. What is left is that Alexander spread Greek culture. See review of Tarn thesis by Eugene Borza, "Ethnicity and Cultural Policy at Alexander's Court," *Makedonika.* 1995, pp. 149–158.

6. For an excellent discussion, see: Frederick M. Hodges, "The Ideal Prepuce in Ancient Greece and Rome: Male Genital Aesthetics and Their Relation to *Lipodermos,* Circumcision, Foreskin Restoration, and the *Kynodesmē,*" *Bulletin of the History of Medicine* 75 (2001): 385–386, 393.
7. Hodges, pp. 375–405, for excellent discussion of issues.
8. James Michener, *The Source* (New York: Random House, 1965), p. 349.
9. *Novum Testamentum Graece.* Alexander Souter, ed. (Oxford: Clarendon Press, 1956).
10. Arnaldo Momigliano, *Alien Wisdom: The Limits of Hellenization* (Cambridge: Cambridge University Press, 1975), p. 98.
11. 3 Mac. 2:30–31p 3:21; 7:10–15, and discussed by Robert G Hall, "Epispasm and the Dating of Ancient Jewish Writings," *Journal for the Study of Pseudepigrapha* 2 (1988): 77.
12. *Antiquities of the Jews,* 12. 241.
13. Soranus, *Gynecology,* Oswei Temkin, trans. (Baltimore: Johns Hopkins University Press, 1956), 2. 34 (p. 107).
14. Dioscorides, *De materia medica,* 2. 82. 2 (Beck trans., p. 127); this and other references are related in good analysis by Hodges, pp. 394–396.
15. *De materia medica,* 4. 153 4 (Beck trans., p. 309).
16. Fiona Grierson, "The Testament of Moses," *Journal for the Study of the Pseudepigrapha* 17/4 (2008): 275.
17. Testament of Moses. 8. Web site: Pseudo Epigrapha, Apographa, and Sacred Writings. http://www.pseudepigrapha.com/pseudepigrapha/assumptionofmoses.html (accessed 12/5/09).
18. Celsus, *De medicina,* 7. 25a (Spencer trans. 3: 421).
19. Ibid.
20. Jody P. Rubin, "Celsus' Decircumcision Operation. Medical and Historical Implications," *Urology* 16/1 (1980): 121–124.
21. Dirk Schultheiss, Michael C. Truss, Christian Stief, and Udo Jonas, "Uncircumcision: A History Review of Preputial Restoration," *Plastic and Reconstructive Surgery* 101/7 (1998): 1990–1998.
22. Ibid., p. 423 for trans. by Spencer.
23. Rubin, pp. 121–124.
24. Pliny, *Natural History,* 24.38. 59: "Non multum a salice vitilium usu distat vitex, foliorum quoque adspectu, nisi odore gratior esset. Graeci lygnon vocant, alian agnon."
25. *De remediis parabilibus* Bk. 1 (14:543 Kühn ed. of Galen).
26. *De materia medica,* 2. 140.
27. Ibid., 3. 34 (Beck trans., p. 200).
28. John Riddle, *Contraception and Abortion from the Ancient World to the Renaissance* (Cambridge, MA: Harvard University Press, 1992), 28–29, 32, 47, 82–86, 88, 89–92, 95, 98, 101–104, 120–123, 126–127, 136–139, 152–156, 160–161.
29. Harat, Zhila Naghibi et al., "Immobilization Effect of *Ruta graveolens* L on Human Sperm: A New Hope for Male Contraception," *Journal of Ethnopharmacology* 115/1 (2008): 38–41.
30. N. A. Khouri and Z. El-Akawi, "Antiandrogenic Activity of Ruta Graveolens L. in Male Albino Rats with Emphasis on Sexual and Aggressive Behavior," *Neuroendocrinology Letters* 26/6 (2005): 823–829.
31. Suwagmani Das, Seema Parveen, Chander Parkush Kundra, and Ben M. J. Pereira, "Reproduction in Male Rats Is Vulnerable to Treatment with Flavonoid-rich Seed Extracts of *Vitex negundo,*" *Phytotherapy Research* 18 (2004): 10.

32. Ibid., p. 12.
33. S. K. Bhargava, "Estrogenic and Pregnancy Interceptory Effects of Flavoids [VI–VII] of *Vitex negundo* L. Seeds in Mice," *Plantes médicinales et phytothérapie* 18 (1984): 74–79.
34. John Riddle, *Eve's Herbs: A History of Contraception and Abortion in the West* (Cambridge, MA: Harvard University Press, 1997), pp. 57–58, 142, 183, 185, and 189.
35. S. K. Bhargava, "Antifertility Effects of the Flavonoids (VI-VII) of *Vitex negundo* L. Seeds in Dogs," *Plantes médinales et phytothérapie* 20/2 (1986): 188–198, quoting p. 195; S. K. Bhargava, "Antiandrogenic Effects of a Flavonoid-rich Fraction of *Vitex negundo* Seeds: A Histological and Biochemical Study in Dogs," *Journal of Ethnopharmacology* 27 (1989): 327–339.
36. Volker Schulz, Rudolf Hänsel, and Varro E. Tyler, *Rational Phytotherapy: A Physician's Guide to Herbal Medicine.* 3rd ed. (Berlin: Springer, 1997), pp. 240–243.
37. Ibid., p. 243.
38. Löw, 3: 491–494, traces its nomenclature among the ancient, premodern, and modern languages of the region.
39. Lise Manniche, *An Ancient Egyptian Herbal* (Austin: University of Texas Press, 1989), p. 155 citing Ebers E 590, E 744, and E 23.
40. Thompson, *DAB*, pp. 220, 247, and 296.
41. JoAnn Scurlock and Burton R. Andersen, trans. and commentary, *Diagnoses in Assyrian and Babylonian Medical Analyses* (Urbana: University of Illinois Press, 2005), 5.58, on p. 111.
42. Ibid. 5.59, on p. 111.
43. David A. Leeming, *The Oxford Companion to World Mythology* (Oxford: Oxford University Press, 2005), p. 325; Michael Grant and John Hazel, ed., *Who's Who in Classical Mythology* (London: Weldenfeld and Nicolson, 1975), p. 345.
44. Galen, *De locis affectis,* 6. 6 (8: 439 Kühn), and translated by Heinrich von Staden, "Spiderwoman and the Chaste Tree: The Semantics of Matter," *Configurations* 1/1 (1993): 30. Priapism's treatment (not cited by von Staden) appears in Galen, *De methodo medendi,* 14. 7 (10: 970).
45. Dioscorides, *De materia medica,* 3. 132 (Beck trans, p. 240).
46. On ancient Vedic medicine see translated text: http://www.archive.org/stream/interpretationof00chakuoft/interpretationof00chakuoft_djvu.txt (accessed 11/27/09).
47. Aelian, *On the Characteristics of Animals,* 9. 26. A. F. Scholfield, trans. in 3 vols. (Cambridge, MA: Harvard University Press; London: Heinemann, 1958–1959) 2:247.
48. von Staden trans., p. 29, based on text, Galen, *De simpl. med. temp. fac.,* 6. 2 (Kühn ed., 11:807–808) with similar information found in Galen, *De alimentorum fac.,* 1. 35 (*Corpus Medicorum Graecorum.* 4. 2 (Helmreich ed., 5./4/2: 259–260); Oribasius. *Coll. Med.* 1.33 (*CMG.* 6/1/1. Raeder ed., p. 18); Aetius of Amida. 1.3 (*CMG* 8/10, Olivieri ed., p. 31) and Paul of Aegina 7. 3. 1 (*CMG* Heiberg, pp. 187, 15–19).
49. Pliny. *Natural History.* 24. 38. 62. W. H. S. Jones' translation (7: 47) reads "violent sexual desire," making a dubious distinction between nonviolent, nonpassionate, or insincere desire. I suspect he was interpreting because he may not have realized that the tree actually can work as Pliny stated.
50. Oribasius, *Ad Eunapium,* 1.35 (*CMG* 6/3, p. 330 12–17 Raeder ed.); von Staden, p. 29.
51. Paul of Aegina, *Seven Books of Medicine,* 7. 3. *s.v. agnos.* (Francis Adams, trans. in 3 vols. 3:20).

52. Galen, *De sanitate tuenda,* 6. 14 (*CMG* 5.4,2, Koch ed., p. 195, lines 32–34); also, see von Staden, p. 28.
53. Dioscorides, *De materia medica,* 1. 103 (Beck trans., p. 74); von Staden, p. 27, prefers to translate the verb, *ek-lyei,* as "release," "set free," or "loosen" rather than Beck's "slacken."
54. Aelian, 9. 26 (von Staden trans., p. 32).
55. "Chaste tree." *Encyclopaedia Britannica.* http://search.eb.com.www.lib.ncsu.edu:2048/eb/article-9022669 (accessed 11/28/2008).
56. Franz-Christian Zygan and Johannes G. Mayer, "Agnus-Castus," in *Die Forschergruppe Klostermedizin hortorische monographien: http://www.klostermedizin.de/index.php?cat=195&cl2=193&cl3=195&art=45* (accessed 12/4/2008). Other German words for the plant are *Schaffmulle* and *Keuschlamm,* whose etymologies are uncertain. See Hermann Fischer, *Mittelalterliche Pflanzenkunde* (Munich: Verlag der Münchner Drucke, 1929), p. 288.
57. *Meyers Konversationslexikon.* Verlag des Bibliographisches Instituts, Leipzig und Wien, 4 parts, 1885–1892 reprinted on Web site: http://www.retrobibliothek.de/retrobib/seite.html?id=100240 (accessed 12/4/09; see *s.v.* vitex).
58. Dioscorides, *De materia medica,* 1: 103.
59. Benedictine Rule 22: http://www.fordham.edu/halsall/source/rul-benedict.html (accessed 12/8/08).
60. Jean Provost, *De remediorum cum simplicium, tum compositione materia. . .* (Venice, 1640), pp. 407–410.
61. E. Mary Smallwood, "The Legislation of Hadrian and Antoninius Pius against Circumcision," *Latomus* 18 (1959): 340.
62. Hodges, pp. 388–392, esp. 390.
63. Amnon Linder, *The Jews in Roman Imperial Legislation* (Detroit: Wayne State University Press, 1987), p. 100; Hodges, pp. 390–392.
64. Pliny, *Natural History,* 24. 38. 59 (W. H. S. Jones, trans. 7: 47).
65. Dioscorides, *De materia medica,* 1. 103 (Beck trans., p. 75).
66. Von Staden, p. 38.
67. Ibid. 41.
68. Christopher Hobbs, *Vitex. The Women's Herb* (Summertown, TN: Healthy Living, 2003; 1st ed., 1990), p. 10; and numerous Web sites, such as http://www.tidesoflife.com/vitex.htm and http://www.herbs2000.com/herbs/herbs_chaste_tree.htm (both accessed 12/8/08). I have been unable to find the quotation in Gerard but, given how much he wrote, it is entirely possible that I missed it. None of the modern authors give a specific citation.
69. John Gerard, *The Herball or Generall Historie of Plants.* Fasc. reproduction of 1597 edition (Amsterdam: Walter J. Johnson, 1974), p. 1202.
70. Hobbs.
71. Ibid., p. 12 (with references).
72. *Diseases of Women.* 1. 44 (Littré ed. 8: 102).
73. Mark Blumenthal, ed. *Herbal Medicine: Expanded Commission E Monographs* (Austin: American Botanical Council, 2000), p. 63.
74. Ibid., p. 62.
75. W. Wuttke, Ch. Gorkow, and H. Jarry, "Dopaminergic Compounds in *Vitex Agnus Castus,*" in *Phytopharmaka in Forschung und klinischer Anwendung.* Dieter Loew and Norbert Rietbrock, eds. (Darmstadt: Steinkopff, 1995), p. 90.
76. I. Gerhard et al., "Mastodynon® bei weblicher Sterlität," *Forschende Komplementärmedizin* 5 (1998): 272–278 (quoting from abstract).

77. Hobbs, pp. 12–19 (quoting p. 13).
78. Von Staden, pp. 26–27 (with references to Hippocratic works).
79. Dioscorides, *De materia medica,* I. 103 (Beck trans., p. 74).
80. Soranus, *Gynecology.* 3. 45 and 46. 3–4 (Temkin trans. with emendations by von Staden, p. 30).
81. Dioscorides, *De materia medica,* 4. 71 cf. 4. 72.
82. See references in Riddle (1997), pp. 57, 142, 183, 185, 189.
83. *Homeric Hymn* 4. *Hymn to Hermes,* 12–16 (Hugh G. Evelyn-White, trans, p. 365).
84. Ibid., 94–96.
85. Ibid., 409–414; on chaste-tree translation, see von Staden, p. 44.
86. Ibid., 491–494 (von Staden trans., p. 44).
87. Ibid., 498 (Evelyn-White trans., p. 399).
88. Line 977 in Rogers ed.

CHAPTER 6

1. An excellent introduction is by Brian P. Copenhaver, *Hermetica* (Cambridge: Cambridge University Press, 1992), pp. xiii–lxi.
2. Garth Fowden, *The Egyptian Hermes: A Historical Approach to the Late Pagan Mind* (Princeton, NJ: Princeton University Press, 1993), esp. p. 29; Copenhaver, p. 93.
3. Horace, *Epistles,* 2. 1. 156 where Horace has Greece instead of Egypt.
4. Ibid., p. 101.
5. Jack Lindsay, *The Origins of Alchemy in Graeco-Roman Egypt* (New York: Barnes and Noble, 1970), p. 38.
6. *Corpus Hermeticum,* 10, 8a (Walter Scott, trans. with Greek, 1: 192–193).
7. *Aesclepius,* 1. 12. (Scott trans. in *CH.* 1: 310–311; slightly different translation by Copenhaver, p. 74).
8. Strobaeus, Excerpt 11. 43 (Scott 1: 433).
9. Fowden, p. 66.
10. John Riddle, "The Pseudo-Hippocratic Dynamidia," *Journal of the History of Biology* 14 (1981): 284.
11. Antoine Faivre, *The Eternal Hermes from Greek God to Alchemical Magus.* Joscelyn Godwin, trans. (Grand Rapids, MI: Phanes Press, 1995), p. 86.
12. *Corpus Hermeticum,* 16. 2 (Scott trans., p. 265).
13. Fowden, p. 81.
14. *Hermetica,* 16. 2 (Copenhaver trans., p. 58).
15. *Stobaei Hermetica,* 6. 18 (Scott trans., p. 419).
16. Faivre, p. 77.
17. *Corpus Hermeticum,* Libellus. 16 (Copenhaver trans., p. 58; slightly different in Scott trans., 1: 265).
18. I have yet to publish my findings about this treatise. Microfilms of around ten manuscripts containing these texts and, with them, the sigils, are now deposited in the Special Collections of D. H. Hill Library, North Carolina State University. A short discussion of the manuscripts along with some of the listings is found in Lynn Thorndike, *A History of Magic and Experimental Science,* 7 vols. (New York: Columbia University Press, 1923), 2:220–221.
19. Faivre, p. 88.
20. *HC. Libellus* 11 (2) (Scott trans. 1: 223).
21. Ibid., 10 (Scott, 1: 195).
22. Ibid., 11 (2) (Scott, 1: 219).

23. Ibid., 13 (Scott, 1: 251).
24. Ibid., 13 (Scott, 1: 253).
25. Ibid., 13 (Scott, 1: 252).
26. Ibid., 13 (Scott, 1: 262–253).
27. As quoted by Thorndike, 6: 439.
28. Joseph Needham, *Science and Civilisation in China*, 7 vols. (Cambridge, UK: University Press, 1954–2000) vol. 5/5: 21–22; 4: 59.
29. Ibid., 5/5: 21.
30. C. J. S. Thompson, *Alchemy and Alchemists* (Mineola, NY: Dover, 2002), pp. 70–71; Stanton J. Linden, "Introduction," *The Mirror of Alchemy* (New York and London: Garland, 1992), pp. xvi–xx.
31. George Sarton, *Introduction to the History of Science*, 3 vols. in 5 pts. (Baltimore: Williams and Wilkins, 1927–1948), 1:xiv.
32. Acharya Prafull Chandra Rây, *A History of Chemistry in Ancient and Medieval India* (Calcutta: Indian Chemical Society, 1956), pp. 158, 166.
33. *Opus Minor*. 314, and translated by Edmund Brehm, "Bacon's Place in the History of Alchemy," *Ambix* 23/1 (1976).
34. Roger Bacon. *On the Errors of Physicians*, Edward Withington, trans. in *Essays on the History of Medicine Presented to Karl Sudhoff* (Freeport, NY: Books for Libraries Press, repr. 1968, from 1924 ed.), p. 143.
35. Ibid., p. 155.
36. Ibid.
37. Ibid., p. 156.
38. Nancy G. Siraisi, *Taddeo Alderotti and His Pupils: Two Generations of Italian Medical Learning* (Princeton, NJ: Princeton University Press, 1981), p. 301.
39. Ibid.
40. Richard P. H. Greenfield, *Traditions of Belief in Late Byzantine Demonology* (Amsterdam: Hakkert, 1988), p. 36.
41. Ibid. 37.
42. Paracelsus, *The Hermetic and Alchemical Writings of Aureolus Philippus Theophrastus Bombast of Hohenheim, called Paracelsus the Great*, 2 vols. (Berkeley, CA: Shambhala, 1976), 2: 330.
43. A good explanation along with references is found in Michael McVaugh, "Quantified Medical Theory and Practice at Fourteenth-Century Montpellier," *Bulletin of the History of Medicine*, 43 (1969): 397–413.
44. Francis B. Brévart, "Between Medicine, Magic, and Religion: Wonder Drugs in German Medico-Pharmaceutical Treatises of the Thirteenth to the Sixteenth Centuries," *Speculum* 83 (2008): 2.
45. Ibid., p. 4.
46. Isidore, *Etymologies*, 17. 9. 45.
47. Francis B. Brévart, "'Mother of all Herbs': The Magical Plant Mugwort (*Artemisia vulgaris* L.) in Medieval German Wonder Drug Literature," in *Er ist ein wol gevriunder man. Essays in Honor of Ernst S. Dick* (Hiledesheim: Olms, 2009), p. 49.
48. Ibid., quoting *Gart der Gesundheit*, 1. 8.
49. *De materia medica*, 3. 24. 3 (Beck trans., p. 190).
50. Brévart (2009), p. 46, who inserts in brackets ("an individual using [in possession of?] mugwort...."
51. Ibid.
52. Ibid., pp. 16–17.

53. Ibid., p. 51.
54. Raphael Patai's translation. Translation in Revised Standard Version translates Lilith as "night hag." A listing of various English translation is on the Web site: http://www.geocities.com/Wellesley/Garden/4240/isaiah.html (accessed 1/17/09).
55. *Dictionary of Deities and Demons in the Bible.* 2nd ed. (Leiden: Brill, 1999), p. 521.
56. *Larousse Encyclopedia of Mythology* (New York: Barnes and Noble, 1994), p. 149.
57. Greenfield, pp. 11–12.
58. Roger Bacon, *Opus Majus,* quoted by Linden, p. 80.
59. Richard Kieckhefer, "Avenging the Blood of Children: Anxiety Over Child Victims and the Origins of the European Witch Trials," in *The Devil, Heresy and Witchcraft in the Middle Ages* (Leiden: Brill, 1998), pp. 91–110.
60. Theodore Rothman, "DeLaguna's Commentary on Hallucinogenic Drugs and Witchcraft in Dioscorides' Materia Medica," *Bulletin of the History of Medicine* 46 (1972): 562–567.
61. For excerpts from primary sources attesting to the "flying ointment," see Hedwig Schlieffer, *Narcotic Plants of the Old World* (Monticello, NY: Lubrecht and Cramer, 1979), pp. 6–21.
62. Weyer, *De praestigiis daemonum*.... Bk. 3, 1563, as translated and cited by Gregory Zilboorg, *The Medical Man and the Witch During the Renaissance* (New York: Cooper Square, 1969), p. 145.
63. Maxim Leonid Weintraub, "The Ungent-Sabbath Paradox: Coma-inducing Solanaceae Plants and the European Witch Phenomenon," Unpubl. thesis, History Department, North Carolina State University, Raleigh, 1996, p. 55.
64. Ibid., p. 36.
65. Heinrich Kramer and James Sprenger, *The Malleus Maleficarum,* 1. 6. Montague Summers, trans. (New York: Dover, 1971), p. 47.
66. Ibid., 1. 6, p. 40.
67. Ibid., 1. 5–6, pp. 40–44.
68. Another wonder or elixir-type drug was soma whose account is well written by William Scott Shelly, *The Elixir: An Alchemical Study of the Ergot Mushrooms* (Notre Dame: Cross Cultural, 1995).
69. John Riddle, *Eve's Herbs: A History of Contraception and Abortion in the West* (Cambridge, MA: Harvard University Press, 1997).
70. Kramer and Sprenger, p. 41.
71. Henry Boguet, *An Examen of Witches Drawn from Various Trials* (New York: Barnes and Noble, 1971), p. 88.
72. Riddle (1997), p. 118, for references.
73. Gunnar Heinsohn and Otto Steiger, *Die Vernichtung der weisen Frauen* ([Erftstadt:] März Verlag, 1st ed. 1985, rev. c. 2005).
74. Margaret Murray, *The Witch-Cult in Western Europe* (London, 1952) and other later publications; Arne Runeberg, *Witches, Demons and Fertility Magic* (Helsinki: Societas Scientiarum Fennica, vol. 14, 1947).
75. *Decretum.* pt. 2, C. 26. q. 5, c. 12 as translated and quoted in *Witchcraft in Europe 1100–1700. A Documentary History.* Alan C. Kors and Edward Peters, ed. (Philadelphia, PA: University of Pennsylvania Press, 1972), p. 29.
76. Peter Biller, *The Measure of Multitude: Population in Medieval Thought* (Oxford: Oxford University Press, 2000), pp. 111–134.
77. Jeffrey Burton Russell, *Witchcraft in the Middle Ages* (Ithaca, NY: Cornell University Press, 1972), p. 83.

78. Heinrich Marzell, *Zauberpflanzen Hexentränke* (Stuttgart: Kosmos, 1963).
79. Margaret Murray, *The Witch-cult in Western Europe; A Study in Anthropology.* (Oxford: Clarendon Press, 1921), p. 170; see also, Thomas G. Benedek, "The Changing Relationship between Midwives and Physicians during the Renaissance," *Bulletin of the History of Medicine* 51 (1977): 550–564.
80. *OED, s.v.* "witch."
81. *OED, s.v.* "witch" (2).
82. Russell, p. 15.
83. Janet Farrar and Gavin Bone, *Progressive Witchcraft* (Franklin Lakes, NJ: Career Press, 2004), p. 18.
84. D'Arch W. Thompson, *A Glossary of Greek Birds* (Hildesheim: Georg Olms, repr. 1966), pp. 262–263.
85. *OED, s.v.* "witchcraft."
86. Also in Ovid, *Metamorphoses,* 5. 534.

Bibliography

Aaronovitch, David, in *The Guardian,* Tuesday June 10, 2003. http://education.guardian.co.uk/higher/comment/story/0,,974440,00.html (accessed 4/16/08).

Abdelgaleil, Samir A. M., A. Moustafa, H. Abdel-Salam, and Mona A. A. Abdel Rasoul, "Bioactivity of Two Major Constituents Isolated from the Essential Oil of *Artemisia judaica* L.," *Biosource Technology* 99/12 (2007): 5947–5950.

Adams, Jad. *Hideous Absinthe: A History of the Devil in a Bottle.* Madison: University of Wisconsin Press, 2004.

Adele Reinhartz, Adele, and Miriam-Simma Walfish. "Conflict and Coexistence in Jewish Interpretation," in *Hagar, Sarah, and Their Children.* Phyllis Tribble and Letty M. Russell, eds. Louisville, KY: John Knox Press, 2006, pp. 107–110.

Adt, Monika, Peter Schmuker, and Iris Müller, "The Role of Atropine in Antiquity and in Anaesthesia," in Atkinson and Boulton, *History of Anaesthesia*, pp. 40–46.

Aelian. *On the Characteristics of Animals.* A. F. Scholfield, trans. in 3 vols. Cambridge, MA: Harvard University Press; London: Heinemann, 1958–1959.

Agathias Scholasticus. *Fragments of Epigrams* in *The Greek Anthology.* W. R. Paton trans. 5 vols. Cambridge, MA: Harvard University Press; London: Heinemann, 1948, vol. 3.

Al-Kindī, see Kindī.

Allen, D. C. "Donne on the Mandrake," *Modern Language Notes* 74/4 (1959): 393–397.

Al-Radi, Selma, "The Ravages of War and the Challenge of Reconstruction," in *The Looting of the Iraq Museum, Baghdad: The Lost Legacy of Ancient Mesopotamia.* Milbry Polk and Angela M. H. Schuster, eds. New York: Harry N. Abrams, 2005, pp. 207–211.

Ambrose. *Hexameron.* John J. Savage, trans. in vol. 42 of *Fathers of the Church.* New York: Fathers of the Church, 1961.

Annus, Amar. *The Standard Babylonian Epic of Anzu.* State Archives of Assyria Cuneiform Text, vol. 3. Helsinki: Neo-Assyrian Text Corpus Project, 2001.

Aprocryphal Lives of Adam and Eve. Brian Murdoch and J. A. Tasoulas, eds. Exeter: Exeter University Press, 2002.

Apuleius, Pseudo-. *Herbarius.* Ernest Howald and Henry Sigerist, eds. in *CML.* vol. 4. Leipzig and Berlin: Teubner, 1927.

Apuleius. *Metamorphoses.* J. Arthur Hanson, trans. in 2 vols. Cambridge, MA: Harvard University Press, 1992 repr.

Aristophanes. *Thesmophoriazusae* in vol. 3. *Work of.* Benjamin Rogers, ed. Cambridge, MA: Harvard University Press, 1950.

Aristotle. *On Sleep and Dreams.* David Gallop, trans. Warminster, UK: Aris and Phillips, 1996.

Arne Runeberg, Arne. *Witches, Demons and\Fertility Magic,* vol. 14. Helsinki: Societas Scientiarum Fennica, 1947.

Arnold, W. N. "Vincent van Goth and the Thujone Connection," *Journal of the American Medical Association* 260/20 (1988): 3042–3044.

Arnott, W. Geoffrey. *Alexis. The Fragments. A Commentary.* Cambridge Classical Texts and Commentaries, vol. 31; Cambridge, UK: Press Syndicate of the University of Cambridge Press, 1996.

Artemisia, see Wright, Colin.

Assante, Julia. "From Whore to Hierodules: The Historiographic Invention of Mesopotamian Female Sex Professionals," in *Ancient Art and Its Historiography.* A. A. Donohue and Mark D. Fullerton, eds. Cambridge: Cambridge University Press, 2003, pp. 13–68.

Atkinson, Richard S., and Thomas B. Boulton, eds. *The History of Anaesthesia.* London: Parthenon, 1987.

Atra-hasī. The Babylonia Story of the Flood. W. G. Lambert and A. R. Millard, trans. with *The Sumerian Flood Story,* M. Civil trans. Oxford: Clarendon Press, 1969.

Attinger, P. "Enki et Ninhursaĝz," *Zeitschrift für Assyriologie und vorderasiatische Archaeologie* 74 (1984): 1–52.

Augustine of Hippo. *Contra Faustum Manichaeum* in: vol. 42 of PL.

Avalos, Hector. *Illness and Health Care in the Ancient Near East: The Rome of the Temple in Greece, Mesopotamia, and Israel.* Atlanta: Scholars Press, 1995.

Avicenna. *Liber Canonis.* Venice ed., 1507 ed.; repr. in fascimle by George Olms, Hildesheim, 1964.

Avigad, Nahman. "The Inscribed Pomegranate from the 'House of the Lord,'" *The Biblical Archaeologist* 53/3 (1990): 157–166.

Ba'al et 'Anat. In *Textes Ougaritiques.,* André Caquot, Maurice Sznycer, and Andrée Herdner, trans. with commentary. Littératures anciennes du Proche-Orient, vol. 7. Paris, Les Éditions du Cerf, 1974.

Bachofen, J. J., *Myth, Religion and Mother Right.* Ralph Manheim, trans. Princeton, NJ: Princeton University Press, 1967.

Bacon. Roger. *On the Errors of Physicians,* Edward Withington, trans. in *Essays on the History of Medicine Presented to Karl Sudhoff.* Freeport, NY: Books for Libraries Press, repr. 1968, from 1924 ed., pp. 139–157.

Bahrani, Zainab. "Performativity and the Image: Narrative, Representation, and the Uruk Vase," in *Leaving No Stones Unturned: Essays on the Ancient Near East and Egypt in Honor of Donald P. Hansen.* Erica Ehrenberg, ed. Winona Lake, IN: Eisenbrauns, 2002, pp. 15–22.

———. *Women of Babylon: Gender and Representation in Mesopotamia.* London: Routledge, 2001.

Baker, Phil. *The Dedalus Book of Absinthe.* Sawtry, UK: Dedalus Books, 2001.

Balz-Cochois, Helgard, *Inanna. Wesenbild und Kult einer unmütterlichen Göttin.* Vol. 4 of Studien zum Verstehen fremder Religionen. Gütterloh: Güttersloher Verlagshaus Gerd Mohan, 1992.

Batto, Bernard Frank. *Studies on Women at Mari.* Baltimore: Johns Hopkins University Press, 1974.

Beavis, Ian C. *Insects and Other Invertebrates in Classical Antiquity.* Exeter, UK: University of Exeter Press, 1988.

BBC h2g2, "Popular Wedding Traditions and Superstitiions," http://www.bbc.co.uk/dna/h2g2/A3383633 (accessed 2/11/08).

Benedek, Thomas G. "The Changing Relationship between Midwives and Physicians during the Renaissance," *Bulletin of the History of Medicine* 51 (1977): 550–564.

Bestiary: Being from an English Version of the Bodleian Library Oxford M. S. Bodley 764: With all the Original Miniatures Reproduced in Facsimile. Woodbride, UK: Boydell and Brewer, 1992, pp. 40–41.

Bhargava, S. K. "Antiandrogenic Effects of a Flavonoid-rich Fraction of *Vitex negundo* Seeds: A Histological and Biochemical Study in Dogs," *Journal of Ethnopharmacolgy* 27 (1989): 327–339.

———. "Antifertility Effects of the Flavonids (VI-VII) of *Vitex negundo* L. Seeds in Dogs," *Plantes médinales et phytothérapie* 20/2 (1986): 188–198.

———. "Estrongenic and Pregnancy Interceptory Effects of Flavoids [VI–VII] of *Vitex negundo* L. Seeds in Mice," *Plantes médinales et phytothérapie* 18 (1984): 74–79.

Bhela-Samhitā. Text with English Translation, Commentary and Critical Notes. K. H. Krishnamurthy, trans.; Priya Vrat Sharma, ed. Varanasi, India: Chaukhambha Visvagharati, 2000.

Biggs, Robert D. "Ergotism and Other Mycotoxicoses in Ancient Mesopotamia," *Aula Orientalis. Revista de estudios del Próximo Oriente Antiguo* 11 (1991): 15–21.

———. *ŠA`.ZI.GA. Ancient Mesopotamian Potency Incantations* (=*Texts from Cuneiform Sources*), vol. 2, Locust Valley, NY: J. J. Augustin, 1967.

Biller, Peter. *The Measure of Multitude: Population in Medieval Thought.* Oxford: Oxford University Press, 2000.

Bleibtreu, Erika. *Die Flora der neuassyrischen Reliefs.* Vienna: Institutes für Orientalistik, 1980.

Blumenthal, Mark. ed. *Herbal Medicine: Expanded Commission E Monographs.* Austin: American Botanical Council, 2000.

Blumer, Dietrich. "The Illness of Vincent van Gogh," *American Journal of Psychiatry* 159/4 (2002): 519–526.

Boasse-Griffithe, Kate. "The Fruit of the Mandrake," in *Fontes Aeque Pontes. Eine Festgabe für Hellmut Brunner.* vol. 5 of *Ägypten und Altes Testament.* Wiesbaden: Otto Harrassowitz, 1983, pp. 62–72.

Bogdanos, Matthew, with William Patrick. *Thieves of Baghdad.* New York: Bloomsbury, 2005.

Boguet, Henry. *An Examen of Witches Drawn from Various Trials.* New York, Barnes and Noble, 1971.

Borca, Federico. "Towns and Marshes in the Ancient World," in *Death and Disease in the Ancient City.* Valerie M. Hope and Eireann Marshall, eds. London and New York: Routledge, 2000, pp. 74–83.

Borza, Eugene. "Ethnicity and Cultural Policy at Alexander's Court," *Makedonika,* 1995, pp. 149–158.

Bosworth, Joseph. *An Anglo-Saxon Dictionary.* London: Oxford University Press, 1954.

Bottéro, Jean. *Everyday Life in Ancient Mesopotamia.* Antonia Nevill, trans. Cambridge: Cambridge University Press, 2001.

Bottéro, Jean. *Mesopotamia, Writing, Reasoning, and The Gods.* Zainab Bahrani and Marc Van de Mieroop, trans. Chicago: University of Chicago Press, 1992.

Bouchez, Colette. Reviewed by Brunilda Nazario, *WebMD*: "Infertility and Reproduction Health Center": http://www.webmd.com/infertility-and-reproduction/features/infertility-stress (accessed 7/24/08).

Brehm, Edmund. "Bacon's Place in the History of Alchemy," *Ambix* 23/1 (1976): 53–58.

Brelich, Angelo. *Paides e Parthenoi.* Rome: Ediziioni dell'Ateneo, 1969.

Brévart, Francis B. "Between Medicine, Magic, and Religion: Wonder Drugs in German Medico-Pharmaceutical Treatises of the Thirteenth to the Sixteenth Centuries," *Speculum* 83 (2008): 1–57.

Brévart, Francis B. "'Mother of all Herbs': The Magical Plant Mugwort (*Artemisia vulgaris* L.) in Medieval German Wonder Drug Literature," in *Er ist ein wol gevriunder man: Essays in Honor of Ernst S. Dick*. Hiledesheim: Olms, 2009, pp. 43–72.

Brooks, Beatrice A. "Fertility Cult Functionaries in the Old Testament," *Journal of Biblical Literature* 60/3 (1941): 227–253.

Brown, Carolyn. "Plants and Trees of Ancient Egypt," *Inscriptions: The Newsletter of the Friends of Egypt Centre* 9 (Dec. 2001), 9–11.

Bryan, Cyril P. *Ancient Egyptian Medicine: The Papyrus Ebers*. Chicago: Ares, 1974.

Budge, Ernest Alfred Wallis. *The Gods of the Egyptians or Studies in Egyptian Mythology*, 2 vols. London: Methuen, 1904; see also, *Syriac Book of Medicines*.

Budin, Stephanie. *The Myth of Sacred Prostitution in Antiquity*. New York: Cambridge University Press, 2008.

Burns, William R. "East Meets West: How China Almost Cured Malaria," *Endeavor* (pre-publication, 2008; accessed via Science Digest, www.sciencedigest.com 8/31/08.

Butenandt, Adolf, and H. Jacobi, "Über die Darstellung eines krystallisierten pflanzliche Tokokininins (Thelykinins) und seing Identifizierung mit dem α-Follikelhormon," *Zeitschrift für physiologische Chemie*, 218 (1933): 104–112.

Caelius Aurelianus. *On Acute Diseases and On Chronic Diseases*. I. E. Drabkin, ed and trans. Chicago: University of Chicago Press, 1950.

Calame, Claude. Derek Collins, and Janice Orion. *Choruses of Young Women in Ancient Greece: Their Morphology, Religious Role, and Social Functions*. Dereck Collins and Janice Orion, trans. Lantham, MD: Rowman and Littlefield, 2001.

Cambridge History of Classical Literature, 2 vols. Cambridge: Cambridge University Press, 1985.

Caner, Ayşe, Mert Döşkaya, Aysu Değirmenci, Hüseyin Can, Şura Baykan, Ahmet Üner, Gülçin Başdemir, et al. "Comparison of the Effects of *Artemisia vulgaris* and *Artemisia absinthium* Growing in Western Anatolia against Trichinellosis (*Trichinella spiralis*) in Rats," *Experimental Parasitology* 1119/1 (2008): 173–179.

Carmody, Francis J. "Physiologus Latinus Version Y," *University of California Publications in Classical Philology* 12 (1944): 95–134.

Cartledge, Paul. *Sparta and Lakonia: A Regional History, 1300–362 BC*. London: Routledge, 2002.

———. *Spartan Reflections*. Berkeley and Los Angeles: University of California Press, 2001.

Casselman, Bill. "Pomegranate: A Naming Error": http://www.billcasselman.com/wording_room/pomegranate.htm (accessed 2/26/08) for Greek weddings.

Chadwick, Robert. *First Civilizations: Ancient Mesopotamia and Ancient Egypt*. 2nd ed. London: Equinox, 2005.

Charlesworth, M. P. "Livia and Tanaquil," *The Classical Review* 41/2 (1927): 55–.

Chiasson, Hélène, André Bélanger, Noubar Bostanian, Charles Vincent, and André Poliquin, "Acaricidal Properties of *Artemisia absinthium* and *Tanacetum vulgare* (Asteraceae) Essential Oils Obtained by Three Methods of Extraction," *Journal of Economic Entomology* 94/1 (2001): 167–171.

Claudian. *De raptu Proserpinae*. David Slavitt, trans. in *Broken Columns: Two Roman Epic Fragments*. Philadelphia: University of Pennsylvania Press, 1997.

Clay, Albert T. *Documents from the Temple Archives of Nippur (=Babylonian Expedition of the University of Pennsylvania)* Series A: Cuneiform Texts. Philadelphia, 1906, vol. 14, no. 163.

Columella. *On Agriculture*, 3 vols. E. S. Forster and Edward H. Heffner trans. in Loeb Classical Library. Cambridge, MA: Harvard University Press; London: Heinemann, 1955, 1993 repr.

Conrad, Barnaby. *Absinthe: History in a Bottle*. San Francisco: Chronicle Books, 1988.

Constantine. *De gradibus.* Munich Bayerishce Staatsbibliothek MS lat. 267.

Cooper, J. S. "Sacred Marriage and Popular Cult in Early Mesopotamia, in *Official Cult and Popular Religion in the Ancient Near East.* E. Matsushima, ed. Heidelberg: Universitätsverlag C. Winter, 1993, pp. 81–96.

Coppens, J. *La connaissance du bien et du mal et le péché du paradis.* Gembloux, 1948.

Corpus Hermetica, see *Hermetica.*

Corpus Scriptorum Latinorum. Theobaldus *Physiologus* on Web site: http://www.forumromanum.org/literature/theobaldus/physiologus.html (accessed 7/30/08).

Crawford, Harriet. *Sumer and the Sumerians.* 2nd ed. Cambridge: Cambridge University Press, 2004.

Cubukcu, B, D. H. Bray, D. C. Warhurst, A. H. Mericli, N. Ozhatay, and G. Sariyan, "*In Vitro* Antimalarial Activity of Crude Extracts and Compounds from *Artemisia abrotanum* L.," *Phytotherapy Research* 4 (1990): 203–204.

Culpeper. Nicholas. *Complete Herbal.* London: W. Foulsham, n.d.

———. *Pharmacopoeia Londinensis or, The London Dispensatory.* London: 1720.

Cunha, Cheston B., and Burke A Cunha, "Brief History of the Clinical Diagnosis of Malaria: From Hippocrates to Osler," *Journal of Vector Borne Disease* 45 (2008): 194–199.

Curley, Michael J. *Physiologus.* Austin: University of Texas Press, 1979.

Das, Suwagman, Seema Parveen, Chander Parkush Kundra, and Ben M. J. Pereira, "Reproduction in Male Rats Is Vulnerable to Treatment with Flavonoid-rich Seed Extracts of *Vitex negundo,*" *Phytotherapy Research* 18 (2004): 8–13.

Davis, Elizabeth Gould. *The First Sex.* New York: G. P. Putnam's Sons, 1971.

Dawson, Warren R. "Studies in Ancient Materia Medica: The Pomegranate," *American Druggist,* December 1925, pp. 22–24.

De rapu Proserpinae, see Rape of Proserpine.

Dean, P. D. G., D. Exley and T. W. Goodwin, "Steroid Oestrogens in Plants: Re-estimations of Oestrone in Pomegranate Seeds," *Phytochemistry* 10 (1971): 2215–2216.

Del Castillo, J., M. Anderson, and G. M. Rubottom, "Marijuana, Absinthe and the Central Nervous System," *Nature* 253 (January 31, 1975): 365–366.

Delaporte, Louis. *Mesopotamia. The Babylonian and Assyrian Civlization.* V. Gordon Childe, trans. New York: Barnes and Noble, rept. 1970.

Delcourt, Marie. *Stérilité mystérieuse & naissances maléfiques dans l'antiquité classique.*Liège: Faculté de Philosophie et Lettres, 1938.

Dictionary of Deities and Demons in the Bible. 2nd ed. Leiden: Brill, 1999.

Dictionary of World Myth. Peter Bently, ed. New York: Facts on File, 1995.

Dio. *Roman History.* Earnest Cary, 9 vols. In Loeb Classical Library. Cambridge, MA: Harvard University Press; London: Heinemann, 1961.

Diogenes Laertius. *Lives of Eminent Philosophers.* With English trans. by R. D. Hicks in 2 vols. Cambridge, MA: Harvard University Press, 1991.

Dioscorides. *De materia medica.* Lily Y. Beck, trans. *Altertums wissenschaftliche Texte und Studien,* vol. 38. Hildesheim: Olms-Weidmann, 2005.

———. *De materia medica.* Max Wellmann, ed., 3 vols. Berlin: Weidmann, 1958.

Dispensatory of the United States of America. 25th. ed. Philadelphia, PA: Lippincott, 1955.

Doman, A. D., P. C. Zuttermeister, and R. Friedman, "The Psychological Impact of Infertility: A Comparison with Patients with Other Medical Conditions," *Journal of Psychosomatic Obstetrics and Gynaecology* 14 (1993 special issue): 45–52.

Douglas, J. D. "Apple," *The New Bible Dictionary.* Grand Rapids, MI: Eerdmans, 1962, pp. 50–51.

Duke, James A. *CRC Handbook of Medicinal Herbs* Boca Raton, FL: CRC Press, 1985.

———. *Duke's Handbook of Medicinal Plants of the Bible.* Boca Raton, FL: CRC Press, 2008.

Duke, Stephen O., Rex N. Paul, and Lee S. Mark, "Terpenoids from the genus Artemisia as Potential Pesticides," in *Biologically Active Natural Products. Symposiuim. American Chemical Society* 194 (1988/3802): 318–334.

Duplais, Pierre A Treatise of the Manufacture and Distillation of Alcoholic Liquors.... M. McKennie, trans. Philadelphia, PA: Carey Baird, 1871, from Traité de la fabrication des liqueurs et de la distillation des alcools...suivi du traité de la fabrication des eaux et boissons gazeuses et de la description complète des opérations nécessaires pour la distillation des alcools. Paris: Gauthier-Villars, 1866.

Duran-Reynals, M. L. *The Fever Bark Tree: The Pageant of Quinine.* Garden City, NY: Doubleday, 1946.

Ebeling, Erich. *Keilschrifttexte aus Assur religiösen.* Ausgrabungen der deutschen Orient-Gesellschaft in Assur, vol. 1. Leipzig: J. C. Hinrichs'sche Buchhandlung, 1919.

Ebers Papyrus. Cyril P. Bryan, trans. *Ancient Egyptian Medicine. The Papyrus Ebers* Chicago: Ares, 1974.

———. H. Joachim, trans. *Papyros Ebers. Das älteste Buch über Heilkunde.* Berlin: Walter de Gruyter, 1973.

Echlin, Kim. *Inanna from the Myth of Ancient Sumer.* Illustrated by Linda Wolfsgruber. Toronto: Groundwood Books, 2003.

Encyclopedia of Occultism and Parapsychology. 2 vols. J. Gordon Melton, ed. Detroit: Gale Research, 2001.

English Poetry (1170–1892). John Matthews Manley, ed. Boston: Ginn, 1907.

Epic of Gilgamesh. N. K. Sandars, trans. New York: Penguin, 1972.

Evans, John M. *Paradise Lost and the Genesis Tradition.* Oxford: Clarendon Press, 1968.

Faivre, Antoine. *The Eternal Hermes from Greek God to Alchemical Magus.* Joscelyn Godwin, trans. Grand Rapids, MI: Phanes Press, 1995.

Farber, Walter. *Beschwörungsrituale an Istar und Dumuzi.* Wiesbaden: Steiner, 1977.

———. *Beschwörungsrituale an Istar und Dumuzi. Attī Ištar ša Harmaša Dumuzi.* Academie der Wissenschaften und der Literatur, vol. 4. Wiesbaden: Steiner, 1977.

Farrar, Janet, and Gavin Bone. *Progressive Witchcraft.* Franklin Lakes, NJ: Career Press, 2004.

Farsit Net; Persian (Iran) Weddings: http://www.farsinet.com/persian_wedding (accessed 2/26/08).

Ferrelljenkins. Worldpress. http://ferrelljenkins.wordpress.com/2008/02/11/israel-issues-a-new-2-shekel-coin/ (accessed 6/8/08).

Fischer, Hermann. *Mittelalterliche Pflanzenkunde.* Munich: Verlag der Münchner Drucke, 1929.

Foley, Helene P. *The Homeric Hymn to Demeter: Translation, Commentary, and Interpretive Essays.* Princeton, NJ: Princeton University Press, 1994.

Fowden, Garth. *The Egyptian Hermes: A Historical Approach to the Late Pagan Mind.* Princeton, NJ: Princeton University Press, 1993.

Frontinus. *Stratagem.* Charles Bennett, trans. Cambridge, MA: Harvard University Press, 1969.

Galen. *De alimentorum facultatibus.* In *CMG.* Helmreich ed., 5./4/2 Leipzig and Berlin, 1923.

———. *De sanitate tuenda.* 6. 14 *CMG* 5.4,2, Koch ed.

———. Various treatises in *Opera omnia,* 22 vols. Thomas Kühn ed. Leipzig, 1821–1833; repr. Hildesheim: Olms, 1964/5.

Geller, Markham J. *Renal and Rectal Disease Texts.* vol. 7 of *MTU.* Berlin: de Gruyter, 2005.

Geoponica. Henry Beckh, ed. Leipzig: Teubner, 1895.

George, A. R. *The Babylonian Gilgamesh Epic.* 2 vols. Oxford: Oxford University Press, 2002.

Gerard, John. *The Herball or Generall Historie of Plants.* Fasc. reproduction of 1597 edition. Amsterdam: Walter J. Johnson, 1974.

Gerard's Herball. The Essence thereof distilled by Marchus Woodward. From edition of Th. Johnson, 1636. New York: Crescent Books, 1985 ed.

Gerhard, I, B. Patke, B. Monga, A. Blank, and C. Gorkow. "Mastodynon® bei weblicher Sterlität," *Forschende Komplementärmedizin* 5 (1998): 272–278.

Ginzberg, Louis. *The Legends of the Jews*, 7 vols. Philadelphia: Jewish Publication Society of America, 1937.

Glover, Terry. R. *Herodotus.* Freeport, NY: Books for Libraries Press, 1969.

Goltz, Dietlinde. *Studies zur altorientalischen und Grieschischen Heilkunde: Therapie—Arzneibereitung—Rezeptstruktur.* Sudhoffs Archiv, vol. 16. Wiesbaden: Franz Steiner Verlag, 1974.

Goodfriend, Elaine Adler. "Prostitution (OT)," *ABD.* 505–510.

Grant, Michael, and John Hazel, ed., *Who's Who in Classical Mythology.* London: Weldenfeld and Nicolson, 1975.

Graves, Robert. *Adam's Rib and Other Anomalous Elements in the Hebrew Creation Myth.* New York: Thomas Yoseloff, 1958.

Graves, Robert, and Raphael Patai, *Hebrew Myths. The Book of Genesis.* Garden City, NY: Doubleday, 1964.

Greene, Mott. T. *Natural Knowledge in Preclassical Antiquity.* Baltimore: Johns Hopkins University Press, 1992.

Greenfield, Richard P. H. *Traditions of Belief in Late Byzantine Demonology.* Amsterdam: Adolf Hakkert, 1983.

Grierson, Fiona. "The Testament of Moses," *Journal for the Study of the Pseudepigrapha* 17/4 (2008): 265–280.

Grieve, M. *A Modern Herbal.* New York: Barnes and Noble, repr. 1996.

Grimm, Jacob. *Teutonic Mythology.* James Stallybrass, trans. in 4 vols. Gloucester, MA: Peter Smith, 1976.

Grmek, Mirko D. *Diseases in the Ancient World.* Mireille Muellner and Leonard Muellner, trans. Baltimore: Johns Hopkins University Press, 1989.

Gujral, M. L., D. R. Varma, and K. N. Sareen, "Oral Contraceptives. Part I: Preliminary Observations on the Antifertility Effect of Some Indigenous Drugs," *Indian Journal of Medical Research* 48 (1960): 46–51.

Gunkel, Hermann. *Genesis* Göttinger Handkommentar zum Alten Testament.; Göttingen: Vandenhoeck and Ruprecht, 1922.

H. J. Woerdenbag and N. Pras. "Analysis and Quality Control of Commercial *Artemisia* Species," in Wright, *Artemisia,* p. 54

Hall, James. *Dictionary of Subjects in Art.* Rev. ed. New York: Harper and Row, 1974.

Hall, Robert G. "Circumcision," in *ABD*:1:1025–1031.

———. "Epispasm and the Dating of Ancient Jewish Writings," *Journal for the Study of Pseudepigrapha* 2 (1988): 71–86.

Hallo, William W. "The Birth of Kings," in *Love and Death in the Ancient Near East Essays in Honor of Marvin H. Pope,* J. H. Marks and R. M. Good, eds. Guildford, CT: Four Quarters, 1987, pp. 45–52.

Hallo, William W. "The Slander Bride," in *Studies Presented to A. Leo Oppenheim, June 7, 1964.* Chicago: University of Chicago Press, 1964, pp. 95–105.

Halloran, Halloran A. Sumerian Lexicon: http://www.sumerian.org/suml-r.htm (accessed 4/6/09).

Handley, M. L. "Quercia," in *Catholic Encyclopedia.* 1911 ed. 12: 601.

Hanuš, Lumír O., Tomáš Řezanka, Jaroslav Spížek, and Valery M. Dembitsky, "Substances Isolated from *Mandragora* Species," *Phytochemisty* 66 (2005): 2408–2416.

Harat, Zhila Naghibi, Mohammad Reza Sadeghi, Hamid Reza Sadeghipour, Mohammed Kamalinejad, and Mohammed Reza Eshriaghian, "Immobilization Effect of *Ruta graveolens* L on Human Sperm: A New Hope for Male Contraception," *Journal of Ethnopharmacolgy* 115/1 (2008): 38–41.

Harper, Donald J. *Early Chinese Medical Literature: The Mawangdui Medical Manuscripts.* London and New York: Kegan Paul, 1998.

Harrar, Sarí, and Sara Altshul O'Donnell. *The Women's Book of Healing Herbs: Healing Teas, Tonics, Supplements, and Formulas.* Emmaus, PA: Rodale Press, 1999.

Harris, Diane, Emily Besselink, and Navindra P. Seeram, "Assessment of Estrogenicity of Pomegrante in an *In Vitro* Bioassay, in Seeram, Schulman, and Heber, *Pomegranates*, pp. 143–156.

Harris, Rivkah. "Independent Women in Ancient Mesopotamia?" in *Women's Earliest Records from Ancient Egypt and Western Asia. Proceedings of the Conference on Women in the Ancient Near East, Brown University, Providence, Rhode Island, November 5–7, 1987* Atlanta: Scholars Press, 1989, pp. 145–165.

Harris, Rivkah. "The *Nadītu* Woman," in *Studies Presented to A. Leo Oppenheim.* Chicago: Oriental Institute, 1964, pp. 106–135.

Harrison, Jane. *Prolegomena to the Study of Greek Religion.* London: Merlin, 1961.

Harrison, R. K. "The Mandrake and the Ancient World," *The Evangelical Quarterly* 28/2 (1956): 87–92.

Hassig, Debra. *Medieval Bestiaries. Text, Image, Ideology.* Cambridge: Cambridge University Press, 1995.

Heftmann, Erich, Shui-Tze Ko, and Raymond D. Bennett, "Identification of Estrone in Pomegranate Seeds," *Phytochemistry* 5 (1966): 1337–1339.

Heidel, Alexander. *Gilgamesh Epic and Old Testament Parallels.* Chicago: University of Chicago Press, 1963.

Heinsohn, Gunnar, and Otto Steiger. *Die Vernichtung der weisen Frauen* [Erftstadt:] März Verlag, 1st ed. 1985, rev. 2005.

Hemingway, Ernest. *For Whom the Bells Toll.* New York: Charles Schribner's, 1940.

Hendel, Ronald S. "Genesis, Book of," in *ABD* 2: 933–941.

Hendin, David. *Guide to Ancient Jewish Coins.* New York: Attic Books, 1976.

Henshaw, Richard A. *Females and Male: The Cultic Personnel: The Bible and the Rest of the Ancient Near East.* Allison Park, PA: Pickwick, 1994.

Hepper, F. Nigel. *Pharaoh's Flowers: The Botanical Treasures of Tutankhamun.* London: HMSO, 1900.

Hermetica. The Ancient Greek and Latin Writings which Contain Religious or Philosophic Teachings Ascribed to Hermes Trismegistus. Trans. and commentary by Walter Scott, in 4 vols. Boston, Shambhala, 1985.

Herodotus. *Persian Wars.* A. D. Godley, trans. Cambridge, MA: Harvard University Press; London: Heinemann, 1990.

———. *Persian Wars.* George Rawlinson, trans. New York: Modern Library, 1942.

Herrewijn, Paul., Adrianna M. van Oosten, and Paul G. M. Piron. *Natural Terpenoids as Messangers: A Multridisciplinary Study of Their Production, Biological Functions, and Practical Applications.* Dordrecth: Kluwer Academic, 2001.

Hesiod. *Theogony.* Hugh Evelyn-White, trans. in Loeb Classical Library. Cambridge, MA: Harvard University Press, 1954.

———. *Work and Days.* 504, H. G. Evelyn-White, trans. Classical Electronic Texts: http://www.theoi.com/Text/HesiodWorksDays.html (accessed 4/26/08).

Hesychius of Alexandria. *Lexikon*, in 5 vols. Amsterdam: Hakkert, 1965.

Hildegard of Bingen. *The Book of Blessed Hildegard begins, Physica*. Bruce W. Hozeski, trans. Boston: Beacon Press, 2002.

———. *Physica,* in *Patrologica Latinus,* vol. 197.

Hippocrates. *Airs, Waters, and Places*. W. H. S. Jones, trans. in Loeb Classical Library. London: William Heinemann; Cambridge, MA: Harvard University Press, 1984 repr., 1:65–138.

———. *Diseases*. Paul Potter, trans. in Loeb Classical Library. Cambridge, MA: Harvard Unitersity Press, 1988, vol. 5: 93–333.

———. *Epidemics* I. W. H. S. Jones, trans. in Loeb Classical Library. London: William Heinemann; Cambridge, MA: Harvard University Press, 1984 repr., 1: 139–211.

———. *Hippocratic Writings*. J. Chadwick, W. N. Mann, I. M. Lonie, and E. T. Withington, trans. : Harmondsworth, UK: Penguin Books, 1978.

———. *Oeuvres completes d'Hippocrate*. Émil Littré, ed. with Fr. trans. in 10 vols. Amsterdam: Hakkert, 1973–1982.

Hobbs, Christopher. *Vitex: The Women's Herb*. Summertown, TN: Healthy Living, 2003; 1st ed., 1990.

Hodges, Frederick M. "The Ideal Prepuce in Ancient Greece and Rome: Male Genital Aesthetics and Their Relation to *Lipodermos,* Circumcision, Foreskin Restoration, and the *Kynodesmē*," *Bulletin of the History of Medicine* 75 (2001): 375–405.

Holtzman, Robert S. "The Legacy of Atropos, the Fate Who Cut the Thread of Life," *Anesthesiology* 89/1 (1998): 241–249.

Homer, see also *Hymn to Demeter*.

———. *Iliad*. A. T. Murray trans. in Loeb Classical Library. Cambridge, MA: Harvard University Press, 1925.

———. *Odyssey,* 2 vols. A. T. Murray, trans. in Loeb Classical Library. Cambridge, MA: Harvard University Press, 1984.

Homeric Hymn. Hymn to Hermes. In *Hesiod, the Homeric Hymns, and Homerica*. Hugh G. Evelyn-White, trans. Cambridge, MA: Harvard University Press, 1954.

Hongwen, Y., and Shouming Zhong, "*Artemisia* Species in Traditional Chinese Medicine and the Discovery of Artemisinin," in Wright, *Artemisia*, pp. 149–157.

Horace. *Odes and Epodes*. David Multoy, ed. and trans. Ann Arbor: University of Michigan Press, 1994.

Hornsey, Ian. S. *A History of Beer and Brewing*. Cambridge, UK: Royal Society of Chemistry, 2003.

Hsu, Elizabeth. "Reflections of the 'Discovery' of the Antimalarial *Qinghao*," *British Journal of Clinical Pharmacology* 61/6 (2006): 666–670.

Huang Di. *The Yellow Emperor's Classic of Medicine*. Maoshing Ni, trans. Boston and London: Shambhala, 1995.

Hunt, Tony. *Plant Names of Medieval England*. Cambridge, MA: D. S. Brewer, 1989.

Hunter, R. L., commentary to: *Apollonius of Rhodes: Argonautica Book III* Cambridge: Cambridge University Press, 1989.

Hymn to Demeter, *The Homeric Hymn to Demeter*. Trans., commentary, and interpretive essays. Helene P. Foley, ed. Princeton, NJ: Princeton University Press, 1994.

Hymn to Demeter. Hugh G. Evelyn-White, trans. from 1914 Loeb Classical Library edition. http://www.sacred-texts.com/cla/demeter.htm (accessed 5/15/08).

I.S.S. [*sic*] *Nuptial Rites, Or the Several Marriage Ceremonies Practised amongst all the Nations in the World* (London: "Printed by T. S. for the Author," 1685), p. 27;

Ibn Sīnā. *Liber Canonis*. Gerard of Cremona, trans. Venice, 1507.

Immerwahr, Sara A. "The Pomegranate Vase: Its Origins and Continuity," *Hesperia* 54/4 (1989): 397–410.

Interlinear Hebrew/Greek English Bible, 4 vols. Jay Green, ed. and trans. Wilmington, DE: Associated Publishers and Authors, 1976.

Internet Drug List: http://www.rxlist.com/cgi/generic/atrop_cp.htm (accessed 7/3/08).

Interpreter's Bible, 12 vols. New York: Abingdon Press, 1954.

Isidore. *Origines* [Etymologies]. Text by W. M. Lindsay, 2 vols. Oxford: Clarendon Press, 1957.

Jackson, Betty P., and Michael I. Berry, "Hydroxytropane Tiglates in the Roots of *Mandragora* Species," *Phytochemistry* 12/5 (1973): 1165–1166.

Jacob, Irene and Walter, "Flora," in *ABD*, 2:803–817.

Jacobsen, Thorkild. *The Harps that Once... Sumerian Poetry in Translation*. New Haven, CT: Yale University Press, 1987.

———. "Pictures and Pictorial Language (The Burney Relief)," in *Figurative Language in the Ancient Near East*. M. Mindlin, M. J. Geller, and J. E. Wansbrough, eds. London: School of Oriental and African Studies, 1987, pp. 5–6.

———. *Toward the Image of Tammuz and Other Essays on Mesopotamian History and Culture*. William L. Moran, ed. Cambridge, MA: Harvard University Press, 1970.

———. *Treasures of Darkness: A History of Mesopotamian Religion*. New Haven, CT: Yale University Press, 1976.

Jacobsthal, Paul. *Greek Pins and Their Connections with Europe and Asia*. Oxford: Clarendon Press, 1956.

Jastrow, Morris. *Religion of Babylonia and Assyria*. Boston: Ginn, 1898.

Jewish Encyclopedia Online: http://www.jewishencyclopedia.com/view.jsp?artid=326&letter=T&search=Twelve%20tribes (accessed 6/24/08).

Johansen, J. Prytz. "The Thesmophora as a Women's Festival," *Temenos* 11 (1975): 80–845.

Joines, Karen Randolf. *Serpent Symbolism in the Old Testament*. Haddonfield, NJ: Haddonfield House, 1974.

Josephus. *Jewish Antiquities*. H. St. J. Thackery, trans. in Loeb Classical Library, vols. 4–9. Cambridge, MA: Harvard University Press, [1926]–1965.

———. *Jewish War*. Gaalya Cornfeld, Banjamin Mazar, and Paul L. Maier, eds. Grand Rapids, MI: Zondervan, 1982.

———. *Jewish Wars*. H. St. J. Thackery, trans. in Loeb Classical Library, vols. 2–3. Cambridge, MA: Harvard University Press; London: Heinemann, [1926]–1965.

———. *New Complete Works of*. Paul Maier based on William Whiston, trans. Grand Rapids, MI: Kregel, 1999.

Juvenal. *The Sixteen Satires*. Peter Green, trans. London: Penguin, 1974.

Kakis, Frederick J. *Drugs: Facts and Fictions*. New York: Franklin Watts, 1982.

Kalantaridou, S. N., A. Makrigiannakis, E. Zoumakis, and G. P. Chrousos, "Stress and the Female Reproductive System," *Journal of Reproductive Immunology* 62/1 (2004): 61–68.

Kamat. S. D. *Studies on Medicinal Plants and Drugs in Dhanvantari-Nighantu*. Delhi: Chaukhamba Sanskrti Pratishthan, 2002.

Kenyon, Kathleen M. *Excavations of Jericho*, 2 vols. London: British School of Archaeology in Jerusalem, 1965.

Khouri, N. A., and Z. El-Akawi, "Antiandrogenic Activity of Ruta Graveolens L. in Male Albino Rats with Emphasis on Sexual and Aggressive Behavior," *Neuroendocrinology Letters* 26/6 (2005): 823–829.

Kieckhefer, Richard. "Avenging the Blood of Children: Anxiety Over Child Victims and the Origins of the European Witch Trials," in *The Devil, Heresy and Witchcraft in the Middle Ages: Essays in Honor of JeffreyB. Russell*. Leiden: Brill, 1998, pp. 91–110.

Kilmer, Anne Draffkown. "The Mesopotamian Concept of Overpopulations and Its Solution as Reflected in the Mythology," *Orientalia* n.s. 41 (1972): 160–177.

Kindī, Al- *The Medical Formulary or Aqrābādhīnī of.* Martin Levy trans. Madison: University of Wisconsin Press, 1966.

King, Leonard W. *A History of Sumer and Akkad.* London: Chatto and Windus, 1916. Klayman, Daniel L., "*Qinghaosu* (Artemisinin): An Antimalarial Drug from China," *Science* n.s. 2228/4703 (1985): 1049–1055.

Köcher, Franz. *Die babylonish-assyrische Medizin in Texten und Untersuchungen,* 7 vols. Berlin: Walter de Gruyter, 1963–2005.

Kramer, Heinrich, and James Sprenger, *The Malleus Maleficarum.*1. 6. Montague Summers, trans. New York: Dover, 1971.

Kramer, Samuel Noah. *History Begins at Sumer: Thirty-nine Facts in Man's Recorded History.* 3rd ed. Philadelphia: University of Pennsylvania Press, 1981.

———. *The Sacred Marriage Rite: Aspects of Faith, Myth, and Ritual in Ancient Sumer.* Bloomington: Indiana University Press, 1969.

Kraft, Carola; Kristina Jenett-Siems, Karsten Siems, Jasmin Jakupovic, Steven Mavi, Ulrich Bienzle, and Eckart Eich. "*In Vitro* Antiplasmodial Evaluation of Medicinal Plants from Zimbabwe," *Phytotherapy Research* 17 (2003): 123–128.

Kupka, Marka, S., C. Dorn, O. Richter, Andreas Schmutzler, Hans van der Ven, and Andrzej Kulczycki, "Stress Relief after Infertility Treatment-spontaneous Conception, Adoption, and Psychological Counselling," *European Journal of Obstetrics and Gynecology Reproductive Biology* 110/2 (2003): 190–195.

Labat, René. *Traité akkadien de diagnostics et pronostics médicaux.* 2 vols. *Collection de travaux de l'Académie Internationale d'histroires des sciences,* No. 7. Paris: Académie Internationale d'Histoire des Sciences; Leiden: Brill, 1951.

Lacroix, Paul. *A History of Prostitution.* Samuel Putnam, trans. from French. New York: Covici, Friede, 1931.

Lambert, Wildred G. "Devotion: The Languages of Religion and Love," in *Figurative Language in the Ancient Near East.* M. Mindlin, M. J. Geller, and J. E. Wansbrough, eds. London: University of London, 1987.

———. "Gilgamesh in Literature and Art: The Second and First Millennia," in *Monsters and Demons in the Ancient and Medieval Worlds.* Mainz: Philipp von Zabern, 1987.

———. "The Gula Hymn of Bullutsa-rabi," *Orientalia* 36 (1967): 105–132.

Lane-Poole, Stanley. *Coins and Medals: Their Place in History and Art.* London: Elliott Stuck, 1885.

Lanier, Doris. *Absinthe: The Cocaine of the Nineteenth Century.* Jefferson, NC: McFarland, 1995.

Lans, Cheryl, Nancy Turner, Yonya Khan, and Gerhard Brauer. "Ethnoveterinary Medicines Used to Treat Endoparasites and Stomach Problems in Pigs and Pets in British Columbia," *Veterinary Parasitology* 148/1–2 (September 2007): 325–340.

Lansky, E., Shubert, S. and Neeman, I, "Pharmacological and Therapeutic Properties of Pomegranate," in *Production, Processing and Marketing of Pomegranate in the Mediterranean Region: Advances in Research and Technology.* Melgarejo-Moreno P., Martínez-Nicolás J. J., and Martínez-Tomé J. eds. Zaragoza : CIHEAM-IAMZ, 2000, pp. 231–235.

Larousse Encyclopedia of Mythology. New York: Barnes and Noble, 1994.

Latham, R. E. *Revised Medieval Latin Word-List.* London: British Academy, 1965.

Leeming, David. A. *The Oxford Companion to World Mythology.* Oxford: Oxford University Press, 2005.

Lehmann, Phyllis Williams. *Statues on Coins of Southern Italy and Sicily in the Classical Period.* New York: H. Bittner, 1946.

Leick, Gwendylyn. *Sex and Eroticism in Mesopotamian Literature.* London: Routledge, 1994.

Lemaire, André. "Probable Head of Priestly Scepter from Solomon's Temple in Jerusalem," *Biblical Archaeology Review* 10 (1984): 24–27.

Lesko, Barbara, ed. *Women's Earliest Records: From Ancient Egypt and Western Asia.* Barbara S. Lesko, ed. Atlanta: Scholars Press, 1989, p. 241.

Lewis, Brian. *The Sargon Legend: A Study of the Akkadian Text and Tale of the Hero Who Was Exposed at Birth.* Cambridge, MA: American School of Oriental Research, 1980.

Life of Adam and Eve. M. D. Johnson, trans. in J. H. Charlesworth, ed. *Old Testament Pseudepigrapha*, 2 vols. Garden City, NY: Doubleday, 1985, 2: 249–295.

Linden, Stanton J. "Introduction," *The Mirrow of Alchemy.* Composed by the Thrice-Famous and Learned Fryer, Roger Bacon. New York and London: Garland, 1992, pp. ix–xlvii.

Linder, Amon. *The Jews in Roman Imperial Legislation.* Detroit: Wayne State University Press, 1987.

Lindsay, Jack. *The Origins of Alchemy in Graeco-Roman Egypt.* New York: Barnes and Noble, 1970.

Liu An. *Divine Farmer's Materia medica: A Translation of the Shen Nong Ben Cao Jing.* Yang Shouzhong, ed. Boulder, CO: Blue Poppy Press, 1998.

Livingstone, Alasdair. *Mystical and Mythological Works of Assryian and Babylonian Scholars.* Oxford: Clarendon Press, 1986.

Löw, Immaneul. *Die Flora der Juden*, 4 vols. Vienna, 1928–1934; Hildesheim: Georg Olms, 1967 repr.

Lucas, Alfred, and J. R. Harris. *Ancient Egyptian Materials and Industries.* London: Arnold, 1962.

Macer. *Herbarum.* [Paris]: Bacquelier, [*ca.* 1515].

Machiavelli, Niccolò. *Mandragola,* David Sices and James B. Atkinson, eds. and trans. Hanover and London: University Press of New England, 1985, pp. 154–275.

Mahmoud, Ahmed, and Ahmed A. Ahmed, "α-Pinene-Type Monoterpenes and Other Constituents from *Artemisia suksdorfii*," *Phytochemistry* 67/19 (October, 2006): 2103–2109.

Maier, Charles S. *Among Empires: American Ascendancy and Its Predecessors.* Cambridge, MA: Harvard University Press, 2006.

Management of Severe Malaria: A Practical Handbook. Geneva: World Health Organization, 2000. Mann, John. *Murder, Magic, and Medicine.* Oxford: Oxford University Press, 1994.

Manniche, Lise. *An Ancient Egyptian Herbal.* Austin: University of Texas Press, 1989.

Marcellus. *De medicamentis liber,* 2 vols. Max Niedermann ed., Jutta Kollesch and Diethard Nickel, trans. in German. *CML.* Berlin: Akademie Verlag, 1968.

Martin, Hubert M. Jr., "Artemis," in *ABD.* 1: 464–465.

Marzell, Heinrich. *Zauberpflanzen Hexentränke.* Stuttgart: Kosmos, 1963.

Mascetti, Manuela Dunn. *Artemis: Goddess of the Hunt and Moon.* San Francisco: Chronicle Books, 1996.

Maspero, Gaston. *A History of Egypt, Caldea, Syria, Babylonia, and Assyria.* 13 vols. London: Gralier Society, [1901?–1906].

Matossian, Mary A. K. *Poisons of the Past: Molds, Epidemics, and History.* New Haven, CT: Yale University Press, 1989.

McCormick, Michael. "Towards a Molecular History of Justinianic Plague," in *Plague and the End of Antiquity: The Pandemic of 541–750.* Lester K. Little, ed. Cambridge: Cambridge University Press, 2007, pp. 290–312.

McDonald, J. Andrew. "Botanical Determination of the Middle Eastern Tree of Life," *Economic Botany* 56/2 (2002): 113–129.

McGregor, I. A. "Malaria," in *The Welcome Trust Illustrated History of Tropical Diseases.* F. E. G. Cox, ed. London: Wellcome Trust, 1996.

McKenzie, John L. "The Literary Characteristics of Genesis 2–3," *Theological Studies* 15 (1954): 541–572.

McVaugh, Michael. "Quantified Medical Theory and Practice at Fourteenth-Century Montpellier," *Bulletin of the History of Medicine* 43 (1969): 397–413.

Meissner, Bruno. *Assyriologische Studien* VI (= *Mitteilungen der vorderasistischen/aegyptischen Gesellschaft* 18/II). Berlin, 1913.

Meschler, J. P., and A. C. Howelett, "Thujone Exhibits Low Affinity for Cannabionoid Receptors but Fails to Evoke Cannabimimetic Responses," *Phmarmacology Biochemistry and Behavior* 62/3 (1999): 473–780.

Meshnick, S. R., T. E. Taylor, and S. Kamchonwongpaisan. "Artemisinin and Antimalarial Endoperoxides: From Herbal Remedy to Targeted Chemotherapy," *Microbiological Reviews* 60/2 (1996): 301–315.

Meyers Konversationslexikon. Verlag des Bibliographisches Instituts, Leipzig und Wien, 4 parts, 1885–1892 reprinted on Web site: http://www.retrobibliothek.de/retrobib/seite.html?id=100240 (accessed 12/4/09).

Meyers, Carol. "Jaachin and Boaz," *ABD* 3: 597–598.

———. "Jachin and Boaz in Religious and Political Perspective," *Catholic Biblical Quarterly* 45/2 (1983): 167–178.

Michener, James. *The Source.* New York: Random House, 1965.

Midrash Rabbah. H. Freedman and Maurice Simon, trans., 5 vols. London: Soncino Press, 1977.

Moldenke, Harold N., and Alma L. Moldenke. *Plants of the Bible.* Waltham, MA: Chronica Botanica Company, 1952.

Momigliano, Arnaldo. *Alien Wisdom: The Limits of Hellenization.* Cambridge: Cambridge University Press, 1975.

Monsters and Demons in the Ancient and Medieval Worlds. Mainz: Philipp von Zabern, 1987.

Müller-Ebeling, Claudia. "Die Alraune in der Bibel," in *Die Sage vom Galgenmännlein im Volksglauben und in der Literatur.* Aldfred Schlosser ed. Berlin: Express Edition, 1987, pp. 141–149.

Mullo-Weir, G. J. "Four Hymns to Gula," *Journal of the Royal Asiatic Society* [no vol. number] January, 1929, 1–18.

Muradyan. *Gohar. Physiologus: The Greek and Armenian Versions with a Study of Translation Technique.* Leuven: Peeters, 2005.

Murray, Margaret. *The Witch-Cult in Western Europe: A Study in Anthropology.* Oxford: Clarendon Press, 1921.

———. *The Witch-Cult in Western Europe. Oxford: Clarendon Press,* 1952.

Murray, Mary Anne. "Fruits, Vegetables, Pulses and Condiments," in *Ancient Egyptian Materials and Technology.* Paul T. Nicholson and Ian Shaw, eds. Cambridge and New York: Cambridge University Press, 2000, pp. 609–655.

Muthmann, Friedrich. *Granatapfel, Symbol des Lebens in der Alten Welt.* [Freibourg:] Office du livre, 1982.

Myres, John L. "Persephone and the Pomegranate (*H. Dem.* 372–4)," *Classical Review* 52/2 (1938): 51–52.

Myth of Adap, see Rogers, Robert.

Namdeo, A. G., K. R. Mahadik, and S. S. Kadam, "Antimalarial Drug—*Artemisia annua*," *Pharmacognosy Magazine* 2 /6 (2006): 106–109.

Nature's Medicines that Heal. Washington: National Geographic Society, 2000.

Needham, Joseph. *Science and Civilisation in China*, 7 vols. Cambridge, UK: University Press, 1954–2000.

Neusner, Jacob. *Comparative Midras: The Plan and Program of Genesis and Leviticus Rabbah.* Atlanta: Scholars Press, 1986.

New York Times, 1 February 2008, A4.

Nicander. *Theriaca* and *Alexipharmaca* in *Nicander: The Poems and Poetical Fragments.* A. S. F. Gow and A. F. Scholfield, eds. Cambridge: Cambridge University Press, 1953.

Norris, Pamela. *Eve: A Biography.* New York: New York University Press, 1999.

Novum Testamentum Graece. Alexander Souter, ed. Oxford: Clarendon Press, 1956.

Nunn, J. F. "The Origins of Anasthesia," in Atkinson and Boulton, *History of Anaesthesia,* pp. 21–26.

Nutton, Vivian. "Medical Thoughts on Urban Pollution," in *Death and Disease in the Ancient City.* Valerie M. Hope and Eireann Marshall, eds. London and New York: Routledge, 2000, pp. 65–73.

O'Brien, Joan. "Nammu, Mani, Eve and Pandora: 'What's in a Name?'?" *Classical Journal* 79/1 (1983): 35–45.

Oppenheim, A. Leo. *Ancient Mesopotamia: Portrait of a Dead Civilization.* Chicago: University of Chicago Press, 1964.

Options Méditerranéennes : Série A. Séminaires *Méditerranéennes*; n. 42). ISBN 2–85352-214–8. Symposium on "Production, processing and marketing of pomegranate in the Mediterranean region: Advances in research and technology." Online version: http://ressources.ciheam.org/om/pdf/a42/00600277.pdf (accessed 4/24/08).

Oribasius. *Ad Eunapium.* Johannes Raeder, ed. in *CMG.* 6/3. Leipzig and Berlin, 1926; repr. Amsterdam: Hakkert, 1964.

Ovid. *Fasti.* James G. Frazer, trans. Cambridge, MA: Harvard University Press; London: Heinemann, 1951.

———. *Metamorphoses.* 2 vols. Frank J. Miller, trans. Loeb Classical Library. London: Heinemann; Cambridge, MA: Harvard University Press, 1916.

Pagels, Elaine. *Adam, Eve, and the Serpent.* New York: Random House, 1988.

Pankaj Oudhia. "Research Note," in http:botanical.com/site/column_poudhis/77_gyn.html (accessed 3/21/08; site Botanical.com is by subscription).

Panyu Tiger. http://panyutiger.blogspot.com/2008/05/chinese-medical-journal-paper-1979.html (accessed 8/31/08).

Paracelsus. *The Hermetic and Alchemical Writings of Aureolus Philippus Theophrastus Bombast of Hohenheim, called Paracelsus the Great,* 2 vols. Berkeley: Shambhala, 1976.

Patai, Raphael, see Geocities under Web sites.

Patai, Raphael. *Sex and Family in the Bible and the Middle East.* Garden City, NY: Putnam's Sons, 1959.

Paul of Aegineta. *Seven Books of Medicine.* Francis Adams, trans. and commentary in 3 vols. London: Sydenham Society, 1847.

Paulus Silentiarius. *Fragments of Epigrams* in *The Greek Anthology.* 5 vols. W. R. Paton trans. Cambridge, MA: Harvard University Press; London: Heinemann, 1948, vol. 3.

Pausanius. *Description of Greece.* W. H. J. Jones, trans. in Loeb Classical Library, 5 vols. Cambridge, MA: Harvard University Press, 1918.

PDRHealth. Physicians' Desk Reference online: http://www.pdrhealth.com/drugs/rx/rx-mono.aspx?contentFileName=Lev1223.html&contentName=Levsin&contentId=302 (accessed 8/1/08).

Pepys, Samuel. *Diary of...* Raleigh: Hayes Barton Press, 2007, repr. of 1660 edition.

Philip Smith, "Mausoleum," in *Dictionary of Greek and Roman Antiquities.* William Smith, ed. London: Murray, 1875, pp. 744–745.

Phillips, John A. *Eve: The History of an Idea.* San Francisco: Harper & Row, 1984.

Philo. *De vitae Mosis.* Bk. 2. 24. 119–121 in *The Works of Philo: Complete and Unabridged.* C. D. Yonge, trans. Peabody, MA: Hendrickson, 1993.

Physiologus, see also Carmody, Theobaldus, and *Corpus Scriptorum Latinorum.*

Physiologus. The Middle English Physiologus. Hanneke Wirtjes, ed. Oxford: Early English Text Society, 1991.

Pirenne-Delforge, Vinciane. "Review of: Stephanie Budin, *The Myth of Sacred Prostitution…*, in *Bryn Mawr Classical Review* (2009). 04. 28 online review, 8 pp.

Platearius. *Circa instans.* in *Das Arzneidrogenbuch Circa instans in einer Fassung des XIII. Jahrhunderts aus der Universitäsbibliothek Erlangen.* Hans Wölfel, ed. dissertation. Berlin: Preilipper, 1939.

Plaut, W. Gunther, Bernard J. Bamberger, and William H. Hallo. *Torah: The Torah, a Modern Commentary.* New York: Union of American Hebrew Congregation, 1981.

Plutarch. *Moralia. Quomodo adolescens poetas audire debeat.* Vol. 1 of Frank Cole Babbit, trans. in Loeb Classical Library, 15 vols. Cambridge, MA: Harvard University Press, 1949.

Pliny (the Elder). *Natural History,*10 vols. in Loeb Classical Library. Cambridge, MA: Harvard University Press; London: Heinemann, 1938–1963.

Polyaenus. *Stratagems of War.* Peter Krentz and Everett L. Wheeler, ed and trans. in 2 vols. Chicago: Ares, 1994.

Pomegranates: Ancient Roots to Modern Medicine. Navindra P. Seeram, Risa N. Schulman, and David Heber, eds. Boca Raton, FL: CRC Taylor & Francis Press, 2006.

Pope, Marvin H. *Song of Songs: A New Translation with Introduction and Commentary.* Anchor Bible Series. Garden City, NY: Doubleday, 1977.

Powell, Marvin A. "Classical Sources and the Problem of the Apricot," *Bulletin on Sumerian Agriculture* 3 (1987): 153–156.

———. "The Tree Section of *ur5(=HAR)-ra=hubullu,*" *Bulletin on Sumerian Agriculture* 3 (1987): 145–151.

———. "Timber Production in Presargonic Lagaš," in *Trees and Timber in Mesopotamia. Vol. 6 of* Bulletin on Sumerian Agriculture," Cambridge, 1992.

Prakash, Anand O. "Potentialities of Some Indigeneous Plants for Antifertility Activity," *International Journal of Crude Drug Research* 24 (1986): 19–24.

Prakash, Anand O., B. Saxena, S. Shukla, R.K. Tewari, S. Mathur, A. Gupta, S. Sharma, and R. Mathur, "Anti-implantation Activity of Some Indigenous Plants in Rats," *Acta Europaea Fertilitas* 16 (1985): 441–448.

Preuss, Julius. *Biblical and Talmudic Medicine.* New York: Sanhedrin Press, 1978.

Pritchard, James B. *The Ancient Near East: An Anthology of Text and Pictures.* Princeton, NJ: Princeton University Press, 1958.

———. *Ancient Near Eastern Texts Relating to the Old Testament.* Princeton, NJ: Princeton University Press, 1969.

Procopius. *Anecdota.* G. A. Williamson, trans. Penguin Pb. 1966.

Propertius, Sextus. *Elegies.* H. E. Bultler, trans. and text in Loeb Classical Library. Cambridge, MA: Harvard University Press; London: William Heinemann, 1952.

Provost. Jean. *De remediorum cum simplicium, tum compositione materia…* Venice, 1640.

Pseudo-Apuleius, see Apuleius.

Qu'ran. *The Holy Qur-an.* Text [Arabic], Translation, and Commentary by Abdullah Yusuf Ali. [Washington, DC: Islamic Propagation Centre International], 1946.

Rahner. Hugo. *Greek Myths and Christian Mystery.* Brian Battershaw, trans. New York: Harper and Row, 1963; German ed., 1957.

Randolph, Charles Brewster. "The Mandrake of the Ancients in Folk-Lore and Medicine," *Proceedings of the American Academy of Arts and Sciences* 40 (1905): 485–537.

Rape of Proserpine in *Broken Columns. Two Roman Epic Fragments. The Accilleid* of Publius Papinus Statius and *The Rape of Proserpine* of Claudius Claudianus. David R. Slavitt, trans. Philadelphia: University of Pennsylvania Press, 1997.

Rappoport, Angelo S. *Ancient Israel: Myths and Legends.* New York: Bonanza Books, 1987.

Räth, Karin, Katja Taxis, Gitta Walz, Christoph H. Gleiter, Shu-Ming Li, and Lutz Heidi, "Pharmacokinetic Study of Artemisinin after Oral Intake of a Traditional Preparation of *Artemisia annua* L. (Annual Wormwood)," *American Journal of Tropical Medicine and Hygiene* 70/2 (2004): 128–132.

Rätsch, Christian. "Einleitung" to: Alfred Schlosser, *Die Sabe vom Galgenmännlein im Volksglauben und in der Literatur.* Berlin: Express Edition, 1987, pp. vii–xxiv.

Rây, Acharya Prafull Chandra. *A History of Chemistry in Ancient and Medieval India.* Calcutta: Indian Chemical Society, 1956.

Rekand, Tiln, and Ilmer Sulg. "Absint og den kunstneriske kreativitet," *Tidsskrift for den Norske laegeforening tidsskrift for praktish medicinem ny reakke* 123//1 (2003): 70–73.

Riddle, John M. "Ancient and Medieval Chemotherapy for Cancer," *Isis* 76/3 (1985): 319–330.

———. "Coins and Contraceptives. The Plant that Made Kyrene Famous." *Celator* 17/12 (2003): 34–35.

———. *Contraception and Abortion from the Ancient World to the Renaissance.* Cambridge, MA: Harvard University Press, 1992.

———. "Die Pseudo-Hippocratic Dynamidia," *Journal of the History of Biology* 14 (1981): 283–297.

———. *Eve's Herbs: A History of Contraception and Abortion in the West.* Cambridge, MA: Harvard University Press, 1997.

———. "Folk Tradition and Folk Medicine: Recognition of Drugs in Classical Antiquity." In *Folklore and Folk Medicines.* John Scarborough, editor. Madison: American Institute of the History of Pharmacy, 1987, pp. 33–61.

———. "Review of: *Pomegranates. Ancient Roots...* In *Fruit Gardener* 39/1 (2007): 26–28.

———. "Women's Medicines in Ancient Jewish Sources: Fertility Enhancers and Inhibiters," in *Disease in Babylonia.* I. L. Finkel and M. J. Geller, eds. Boston: Leiden, 2007.

Riddle, John M., and J. Worth Estes. "Oral Contraceptives in Ancient and Medieval Times," *American Scientist* 80 (1992): 226–233.

Ro, Dae-Kyun, Paradise, Eric M., Ouellet, Mario, Fisher, Karl J., Newman, Karyn L., Ndungu, John M., Ho, Kimberly A., et al. "Production of Antimalarial Drug Precursor Artemisinic Acid in Engineered Yeast," *Nature* 440 (13 April, 2006): 940–943.

Roger Bacon, *Opus Minor,* see Brehm.

Rogers, Robert Williams. *Cuneiform Parallels to the Old Testament.* New York: Abington Press, 1912.; also via Web site: http://www.sacred-texts.com/ane/adapa.htm (accessed 5/5/08).

Roth, Martha T. "Marriage and Matrimonial Prestations in First Millennium B.C. Babylonia," in Lesko, *Women's Earliest Records,* pp. 245–260.

Rothman, Theodore. "DeLaguna's Commentary on Hallucinogenic Drugs and Witchcraft in Dioscorides' Materia Medica," *Bulletin of the History of Medicine* 46 (1972): 562–567.

Rowling. J. K. *Harry Potter and the Chamber of Secrets.* New York: Scholastics, 1999.

Rubin, Jody P. "Celsus' Decircumcision Operation: Medical and Historical Implications," *Urology* 16/1 (1980): 121–124.

Rufinus. *Herbal* in *The Herbal of Rufinus.* Lynn Thorndike and Francis Benjamin, eds. Chicago: University of Chicago Press, 1946.

Russell, J. B. *Satan: The Early Christian Tradition.* Ithaca, NY: Cornell University Press, 1981.

———. *Witchcraft in the Middle Ages.* Ithaca, NY: Cornell University Press, 1972.

Saggs, H. W. F., *The Might That Was Assyria*. London: Sidgwick and Jackson, 1984.

Sallares, Robert. *Malaria and Rome: A History of Malaria in Ancient Italy*. Oxford: Oxford University Press, 2002.

Salmon, William. *Medicine Practica or Practcal Physick*. London: Boney, 1692.

Sarton, George. *Introduction to the History of Science*, 3 vols. in 5 pts. Baltimore: Williams and Wilkins,1927–1948.

Satomi, H., K. Umemura, A. Ueno, T. Hatano, T. Okuda, and T. Noro. "Carbonic Anhydrase Inhibitors from the Pericarps of *Punica granatum* L.," *Biological and Pharmaceutical Bulletin* 16 (1993): 787–790 [through Medline].

Scarborough, John. "Drugs and Drug Lore in the Time of Theophrastus: Folklore, Magic, Botany, Philosophy and the Rootcutters," *Acta Classica* 69 (2006): 1–29.

———. "Mandrake in Ancient Surgery," unpublished paper presented at the Society for Ancient Medicine jointly with the American Philological Association, January 7, 2006, Montreal, Quebec.

———. *Medical Terminologies: Classical Origins*. Norman: University of Oklahoma Press, 1992.

———. "The Opium Poppy in Hellenistic and Roman Medicine," in *Drugs and Narcotics in History*. Roy Porter and Mikuláš Teich, eds. Cambridge: Cambridge University Press, 1995, pp. 4–23.

Schaeffer, Claude F. "Les fouilles de Ras-Shamra cinquième campagene (printemps 1933)," *Syria* 15 (1934): 105–134.

Schlieffer, Hedwig. *Narcotic Plants of the Old World*. Monticello, NY: Lubrecht and Cramer, 1979.

Schlosser, Alfred. *Die Sabe vom Galgenmännlein im Volksglauben und in der Literatur*. Berlin: Express Edition, 1987 repr.

Schmandt-Besserat, Denise. *When Writing Met Art: From Symbol to Story*. Austin: University of Texas Press, 2007.

Schmidbauer, Wolfgang. "Die magische Mandragora," *Antalos* 10 (1969): 274–286.

Schneider, Wolfgang. *Lexikon zur Arzneimittelgeschicthe*, 7 vols. in 9 parts. Frankfurt a. M.: Govi-Verlag, 1974.

Schultheiss, Dirk, Michael C. Truss, Christian Stief, and Udo Jonas. "Uncircumcision: A History Review of Preputial Restoration," *Plastic and Reconstructive Surgery* 101/7 (1998): 1990–1998.

Schulz, Volker, Rudolf Hänsel, and Varro E. Tyler., *Rational Phytotherapy: A Physician's Guide to Herbal Medicine*. 3rd ed. Berlin: Springer, 1997.

Scott, R. B. Y. "The Pillars Jachin and Boaz," *Journal of Biblical Literature* 58/2 (1939): 145.

Scribonius Largus. *Compositiones*. George Helmreich, ed. Leipzig: Teubner, 1887.

Scullion, John J. "Genesis, The Narrative of," in *ABD* 2: 941–956.

Scurlock JoAnn, Burton R. Andersen, trans. and commentary. *Diagnoses in Assyrian and Babylonian Medical Analyses*. Urbana: University of Illinois Press, 2005.

Segura, J. J., L. H. Morales-Ramos, J. Verde-Star, and D. Guerra, "Inhibición del orecimiento de Entamoeba histolytica y. E. invandens producida por la raiz del granado (*Punica granatum* L.)," *Archivos Investigación Médica* 21 (1990): 235–239.

Seid, U., "Inanna/Ištar, In der Bildkunst," in *Reallexikon der Assyriologie und vorderasiatischen Archäologie*. Berlin: Walter de Gruyter, 1980, 5: 87–89.

Septuagint with Apocrypha: Greek and English. Sir Lancelot C. Brenton, ed. Grand Rapids, Michigan: Zondervan repr. of 1851 ed.

Shanks, Hershel. "Pomegranate Sole Relic From Solomon's Temple, Smuggled Out of Israel, Now Recovered," *Moment. Magazine of Jewish Culture and Opinion* (December 1988), pp. 36–43.

Shelley, William Scott. *The Elixir: An Alchemical Study of the Ergot Mushrooms.* Notre Dame: Crosse Cultural, 1995.

Siraisi, Nancy G. *Taddeo Alderotti and His Pupils: Two Generations of Italian Medical Learning.* Princeton, NJ: Princeton University Press, 1981.

Skinner, John. *A Critical and Exegetical Commentary on Genesis.* 2nd ed. Edinburgh: T. and T. Clark, 1976 repr., [1930 2nd].

———. *Myths and Legends of Flowers, Trees, Fruits, and Plant in All Ages and All Climes.* Philadelphia, PA: J. B. Lippincott, 1925.

Smallwood, E, Mary. "The Legislation of Hadrian and Antonius Pius against Circumcision," *Latomus* 18 (1959): 334–347.

Smith, Philip. "Mausoleum," in *Dictionary of Greek and Roman Antiquities.* William Smith, ed. London: Murray, 1875, pp. 744–745.

Smith, Phillip E. M. "Absinthe Attacks," *Neurology and Art* 6 (2006): 376–381.

Smith, Wesley D. "Fever Pathology in *Epidemics* 5 and 7," in *Theories of Fever from Antiquity to the Enlightenmen.* W. F. Bynum and Vivian Nutton, eds. London: Wellcome Institute, 1981, pp. 1–18.

Soranus. *Gynecology.* Oswei Temkin, trans. Baltimore: Johns Hopkins University Press, 1956.

———. *Soranii Gynaeciorum libri IV. De signis facturarum. De fascii. Vita Hippocratis secundum Soranum.* Johassnes Ilberg, ed. Leipzig and Berlin: Teubner, 1927.

Speiser, E. A. *The Anchor Bible Genesis.* Garden City, NY: Doubleday, 1964.

Spence, Lewis. *Myths and Legends of Ancient Egypt.* New York: Farrar and Reinhart, 1933.

Stannard, Jerry. "The Plant Called Moly," in *Herbs and Herbalism in the Middle Ages and Renaissance.* Katherine Stannard and Richard Kay, eds. Aldershot: Ashgate Variorum, 1999, pp. 256–307.

Stanton, Annett L., Sharon Sears, Marci Lobel, and Robyn Stein DeLuca, "Psychosocial Aspects of Selected Issues in Women's Reproductive Health: Current Status and Future Directions," *Journal of Consulting and Clinical Psychology* 70/3 (2002): 751–770.

Stoll, Marten. *Birth in Babylonia and the Bible: Its Mediterranean Setting.* Cuneiform Monographs 14. Groningen: Styx, 2000.

Stone, Merlin. *When God Was Woman.* New York: Dial Press, 1976.

Stuckey,Johanna H. "Sacred Prostitutes," *MatriFocus: Cross-Quarterly for Goddess Woman Samhain* (Vol. 5–1): http:www.matrifocus.com/SAM05/spotlight.htm (accessed 4/6/09).

Suetonius. *Life of Claudius.* In vol. 2 of *Lives of Caesars,* 2 vols LJ. C. Rolfe, trans. In Loeb Classical Library. Cambridge, MA: Harvard University Press; London: Heinemann, 1950.

Sumerian Lexicon: http://www.sumerian.org/suml-r.htm (accessed 4/6/09).

Sushruta Samhita. Kaviraj Kunjalal Bhishagratna, ed. and trans. in 3 vols. Varanasi: Chowkhamba Sanskriti Series, 1963.

Suter, Ann. *The Narcissus and the Pomegranate: An Archaeology of the Homeric* Hymn to Demeter. Ann Arbor: University of Michigan Press, 2002.

Suwagmani Das, Seema Parveen, Chander Parkush Kundra, and Ben M. J. Pereira, "Reproduction in Male Rats Is Vulnerable to Treatment with Flavonoid-rich Seed Extracts of *Vitex negundo,*" *Phytotherapy Research* 18 (2004): 8–13.

Syriac Book of Medicines. Ernest A. Wallis Budge, ed. and trans., 2 vols. St. Heller: Armorica Book, 1976.

Taberner, Peter V. *Aphrodisiacs: The Science and the Myth.* Philadelphia: University of Pennsylvania Press, 1985.

Tacitus. *Annals.* John Jackson, trans. in Loeb Classical Library. Cambridge, MA: Harvard University Press and London: Heinemann, 1937.

Talmud (Bablylonian). I. Epstein, trans. London: Soncino Press, [1961].

Tan, R. X., W. F. Zheng, and H. Q. Tang, "Biologically Active Substances from the Genus *Artemisia*," *Planta Medica* 64 (1998): 295–302.

Tarn, W. W. *Alexander the Great: Sources and Studies.* Cambridge: Cambridge University Press, 2003.

Testament of Moses. Web site: Pseudo Epigrapha, Apographa, and Sacred Writings. http://www.pseudepigrapha.com/pseudepigrapha/assumptionofmoses.html (accessed 12/5/09).

Teubal, Savina J. *Sarah the Priestess: The First Matriarch of Genesis.* Athens, OH: Swallow Press, 1984.

Theobaldi Physiologus. P. T. Eden, trans. and ed. Leiden: Brill, 1972.

Theophrastus. *Enquiry into Plants.* Arthur Hort, trans. in 2 vols. Cambridge, MA: Harvard University Press, 1980.

Therapeutic Drugs. 2nd ed., in 2 vols. Colin Dollery, ed. Edinburgh: Churchill Livingston, 1999.

Thompson, C. J. S. *Alchemy and Alchemists.* Mineola, NY: Dover, 2002.

———. *The Mystic Mandrake.* New Hyde Park, NY: University Books, 1968.

Thompson, D'Arch W. *A Glossary of Greek Birds.* London and Oxford: Oxford University Press, 1966; repr. Hildesheim: Georg Olms, 1966.

Thompson, R. Campell. "Assyrian Medical Texts," *Proceedings of the Royal Society of Medicine* 19 (1926): 29–76.

———. *Dictionary of Assyrian Botany* [*DAB*]. London: British Academy, 1949.

———. "The Migration of Assyrian Plant-names into the West," *The Classical Review* 38 (1924): 148–149.

Thorndike, Lynn. *A History of Magic and Experimental Science,* 7 vols. New York: Columbia University Press, 1923.

Thureau-Dangin, F., "Un Hymne à Išta de lat haute Époque Babylonienne," *Revue d'assyriologie et d'archéologie* 22 (1925): 169–177.

Tides of Life: http://www.tidesoflife.com/vitex.htm (accessed 12/8/08); Herbs 2000: http://www.herbs2000.com/herbs/herbs_chaste_tree.htm (both accessed 12/8/08).

Tigay, Jeffrey H. *The Evolution of the Gilgamesh Epic.* Philadelphia: University of Pennsylvania Press, 1982.

Timbrell, John. *The Poison Paradox.* Oxford: Oxford University Press, 2005.

Toynbee, J. M. C. "A New Roman Mosaic Pavement Found in Dorset," *Journal of Roman Studies* 54 (1964): plate 2.

Tribble, Phyllis. and Letty M. Russell, eds. *Hagar, Sarah, and Their Children.* Louisville, KY: Westminster John Knox Press, 2006.

Tusser, Thomas. *Five Hundred Points of Good Husbandry. . . .* William Maver, ed. London: Lackington, Allen, 1812.

Tyler, Varro E. *The Honest Herbal: A Sensible Guide to the Use of Herbs and Related Remedies.* 3rd ed. New York: Pharmaceutical Products Press, 1993.

United States Department of Agriculture. Obnoxious Weeds: http://plants.usda.gov/java/invasiveOne?startChar=A (accessed 8/25/08).

Update. Finds or Fakes; Ivory Pomegranate, *Biblical Archaeology Review* 31/2 (2005), 62–63.

Valecha, Neena, S. Biswas, V. Badoni, KS. Bhandari, O. P. Sati. "Antimalarial Activity of *Artemisia Japonica, Artemisia Maritima, and Artemisia Nilegaricia,*" *Indian Journal of Pharmacology* 26 (1994): 144–146.

Van Arsdall, Anne. "Exploring what was understood by 'mandragora' in Anglo-Saxon England," in *Old Names, New Growth: Proceedings of the 2nd Anglo-Plant Name Survey Symposium*. Graz, June 6–10, 2007. P. Bierbaumer and H.W. Klug eds. Frankfurt/Main: Lang, 2009, pp. 57–74.

Van De Mieroop, Marc. "Women in the Economy of Sumer," in Lesko, *Women's Earliest Records*, pp. 53–69 (incl. discussion of paper).

Van der Toorn, Karel. "Prostitution (Cultic)," *ABD.* 5:510–513.

Varro, Marcus Terentius. *Rerum rusticarum.* 1. 12. 2 William D. Hooper and Harrison B. Ash, trans., in Loeb Classical Library. Cambridge, MA: Harvard University Press; London: William Heinemann, 1979.

———. *Rerum rusticarum.* William D. Hooper and Harrison B. Ash, trans. Cambridge, MA: Harvard University Press, 1934.

von Deines, Hildegard, Hermann Grapow, and Wolfhart Westendorf. *Grundriss der Medizin der alten Ägypter*, 9 vols. Berlin: Akademie Verlag, 1954–1973.

von Soden, W., *Akkadisches Handwörterbuch.* Wiesbaden: Otto Harrassowitz, 1981.

von Staden, Heinrich. "Spiderwoman and the Chaste Tree: The Semantics of Matter," *Configurations* 1/1 (1993): 23–56.

Vulpe, Nicola. "Irony and the Unity of the *Gilgamesh Epic*," *Journal of Near Eastern Studies* 53/4 (1994): 275–283.

Wakeman, Mary K. "Ancient Sumer and the Women's Movement," *Journal of Feminist Studies in Religion* 1/2 (1985): 7–27.

Walafrid Strabo. *Hortulus.* Raef Payne, trans. Pittsburgh: Hunt Botanical Library, 1966.

Wallace, Howard N. "Eve," in *ABD* 2: 676–677.

———. "Adam," in *ABD.* 1: 62–64.

Warren, Michelle P. Eliza Ng, Russalind H. Ramos, and Sari Halpert. "Absence of Significant Estrogenic Effects in the Postmenopausal Population," in Seeram, Schulman, and Heber, *Pomegranates*, pp. 157–164.

Weintraub, Maxim Leonid. "The Ungent-Sabbath Paradox: Coma-inducing Solaraceae Plants and the European Witch Phenomenon," Unpubl. thesis, History Department, North Carolina State University, Raleigh, 1996.

Wellmann, Max. "Allgemeiner Charakter des Physiologos. Zeit und Ort seiner Entstehund," in *Philologus. Zeitschrift für das klassische Altertum,* Suppl.22 (Leipzig: Dieterich'sche Verlagsbuchhandlung, 1930), 22/1: 1–116.

Wernicke, "Artemis," in *Pauys Real-encyclopädie der classischen Altertumswissenschaft* (Stuttgart: Metzler, 1903–) 2/1: 1335–1440.

Werr, Lamia Al-Gailani. "A Museum Is Born," in: *The Looting of the Iraq Museum, Baghdad, The Lost Legacy of Ancient Mesopotamia.* Milbry Polk and Angela M. H. Schuster, eds. New York: Harry N. Abrams, 2005, pp. 27–33

Westermann, Claus. *Genesis 1–11. A Commentary.* John J. Scullion, trans. Minneapolis: Augsburg, 1984 [English ed; 1974 German ed.].Weyer, *De praestigiis daemonum. . . . See Zilboorg.* Bk. 3, 1563, as translated and cited by Mark S. Micale and Roy Porter. *Discoverying the History of Psychiatry.* Oxford: Oxford University Press, 1994, p. 62.

Wheeler, Stephen M. "The Underworld Opening of Claudian's *De Raptu Proserpinae*," *Transactions of the American Philological Association* 125 (1995): 113–134.

White, T. H., ed. *The Bestiary. A Book of Beasts being from a Translation from a Latin Bestiary of the Twelfth Century.* New York: Putnam's Sons, 1960.

Wicke., C., "Inanna/Ištar," in *Reallexikon der Assyriologie und vorderasiatischen Archäologie.* Berlin: Walter de Gruyter, 1980, 5: 74–87.

Widengren, Geo. *The King and Tree of Life in Ancient Near Eastern Religion.* Uppsala: A-B. Lundequistskea Bokhandeln; Leipzig: Harrassowitz. 1951.

Wilkinson, Richard H. *Complete Gods and Goddesses of Ancient Egypt.* London: Thames and Hudson, 1988.

Willcox, Merlin, Gerard Bodeker, Genevidve Bourdy, Vikas Dhingra, Jacques Falquet, Jorge S. Ferreira, Bertrand Graz, et al. "*Artemisia annua* as a Traditional Herbal Antimalarial," in *Traditional Medicinal Plants and Malaria*. Merlin Willcox, Gerard Bodeker, and Philippe Rasanaro, eds. Boca Raton, FL: CRC Press, 2004, pp. 43–60.

Willie Martin Studies. http://www.israelect.com/reference/WillieMartin/The_Story_Of_Judah.htm (accessed 6/14/08).

Wink, Michael. "A Short History of Alkaloids," in *Alkaloids. Biochemistry, Ecology, and Medicinal Aplications.* Margaret F. Roberts and Michael Wink, eds. New York and London: Plenum Press, 1998, pp. 11–44.

Wirtjes, Hanneke. *The Middle English Physiologus.* Oxford: Early English Text Society, 1991.

Witchcraft in Europe 1100–1700: A Documentary History. Alan C. Kors and Edward Peters, eds. Philadelphia: University of Pennsylvania Press, 1972.

Woerdenbag H. J. and N. Pras. "Analysis and Quality Control of Commercial *Artemisia* Species," In Wright, *Artemisia,* pp. 51–77.

Wolkstein, Diane, and Samuel Noah Kramer. *Inanna Queen of Heaven and Earth: Her Stories and Hymns from Sumer.* New York: Harper and Row, 1983.

Wright, Colin W. ed. *Artemisia*. London and New York: Taylor and Francis, 2002.

Wuttke, W., Ch. Gorkow, and H. Jarry, "Dopaminergic Compounds in *Vitex Agnus Castus,*" in *Phytopharmaka in Forschung und klinischer Anwendung.* Dieter Loew and Norbert Rietbrock, eds. Darmstadt: Steinkopff, 1995, pp. 81–91.

Xenophone. *Anabasis.* Carleton Brownstone, trans. in vol. 1 of Xenophone. Loeb Classical Library. Cambridge, MA: Harvard University Press, 1947.

Yin Yang Mai Sihou, Donald J. Harper, trans. in *Early Chinese Medical Literature: The Mawangdui Medical Manuscripts.* London and New York: Kegan Paul, 1998, pp. 219–304.

Yu, Hongwe, and Shouming Zhong. "*Artemisia* Species in Traditional Chinese Medicine and the Discovery of Artemisinin," in Wright, *Artemisia*, pp. 149–157.

Zagarell, Allan, "Responses to Prof. Kuhrt's Paper," in Lesko, see above.

Zilboorg, Gregory. *The Medical Man and the Witch during the Renaissance.* New York: Cooper Square, 1969.

Zohary, David, and Maria Hopf. *Domestication of Plants in the Old World.* 3rd ed. Oxford: Oxford University Press, 2000.

Zohary, Michael. *Plants of the Bible.* Cambridge: Cambridge University Press, 1982.

Zygan Franz-Christian, and Johannes G. Mayer. "Agnus-Castus," in *Die Forschergruppe Klostermedizin hortorische monographien: http://www.klostermedizin.de/index.php?cat=195&cl2=193&cl3=195&art=45* (accessed 12/4/80).

WEB SITES

AccuHist: http://www.accuhist.com/ (accessed 8/1/08).

BBC News, Feb. 8, 2003: http://newsvote.bbc.co.uk/mpapps/pagetools/print/news.bbc.co.uk/2.html (accessed 8/25/08).

Benedictine Rule 22: http://www.fordham.edu/halsall/source/rul-benedict.html (accessed 12/08/08).

Dream Pharmacy http://dreampharm.com/garlic/mugwort.asp (accessed 9/7/08).

EDrugDigest: http://www.drugdigest.org/DD/DVH/HerbsWho/0,3923,4064|Wormwood,00.html (accessed 10/5/08).

Geocities: http://www.geocities.com/heartland/bluffs/8336/robertservice/shooting.html (accessed 10/10/2008).

Geocities: http://www.geocities.com/Wellesley/Garden/4240/isaiah.html (accessed 1/17/09).

Globosapiens: Iranan Wedding Ceremonies: http://www.globosapiens.net/travel-information/Esfahan-1974.html (accessed 2/26/08).

Internet Drug List: http://www.rxlist.com/cgi/generic/atrop_cp.htm (accessed 7/3/08).

Microbiologybytes: http://www.microbiologybytes.com/introduction/Malaria.html (accessed 8/26/08).

Microbiologybytes Web site: BBC h2g2, "Popular Wedding Traditions and Superstitiions," http://www.bbc.co.uk/dna/h2g2/A3383633 (accessed 2/11/08).

Mscl.: Plants. United States Department of Agriculture. http://plants.usda.gov/java/invasiveOne?startChar=A (accessed 8/25/08).

MS ICE: http://www.pmsice.com/Ingredients.cfm (accessed 9/7/08).

Panyu Tiger: http://panyutiger.blogspot.com/2008/05/chinese-medical-journal-paper-1979.html (accessed 8/31/08).

Vedic medicine: http://www.archive.org/stream/interpretationof00chakuoft/interpretationof00chakuoft_djvu.txt (accessed 11/27/09).

Virtual Absinthe Museum http://oxygenee.com/absinthe-effects.html (accessed 10/21/08).

Yahoo. Health: http://health.yahoo.com/flu-medications/atropine-chlorpheniramine-hyoscyamine-phenylephrine-scopolamine/healthwise—d04872a1.html (accessed 8/1/08).

INDEX

Aaron, 40
abortifacient, *see* contraception and abortion
Abraham, 57–58, 76
absinthe, *see* artemisia
AccuHist, 66
Adam, 6, 14, 22–23, 33, 36–40, 72–73, 136–137, 139, 141, 147
Adapa and the Flood of Life, 35
Aesclepius, 131
Aesculapius (alt. Asclepius), 91, 132
Agathias Scholasticus, 99
Agnolyt, 124
agricultural revolution, 1–2
Akkadians and Akkadia, 15, 29, 36, 47–53
alchemy, 134–137
alcohol, 102–111, *see also* beer, wine
Alexander Jannaeus, *see* Yehonatah
Alexander the Great, 114
Alexis, 71
Al-Kindī, 86, 96
Ambrose (Saint), 37
Amenhotep II, 44, 49
American mandrake, 58
anesthesia, 67–68
animal husbandry, 126–127
animal pharmacognosy, 2
Antiochus, 115
Antonius Pius, 122
Anzû bird/*Epic of Anzu*, 14–16, 26, 147
aphrodisiacs, 65–66, 68–70, 75
Aphrodite, 20, 26, 51, 60, 99, 120, 129
Apollo, 79–80, 91, 125–126
apple, 20, 40, 65, 72–73, 139, 141, *see also* love apples, pomegranate
apricot, 40
Apuleius, 71, *see also* Pseudo-Apuleius
Aristophanes, 46, 126
Aristotle, 71, 134
arsenogonon, 70
Artapan, 132
Artemis (goddess), 79–80, 103, 111, 129, 134, 138, 147
Artemis Orthia (festible of), 80
artemisia (plant), 79–112, 129, 134, 138–139, 143, 147: *Artemisia afra*, 95; absinthe (*A. absinthium),* 81–82, 84–88, 95–97, 103–111; annual wormwood (*A. annua*), 81, 93–95, 98; *A. apiacea*, 93–94; field mugwort (*A. arborescens*), 81, 83; *A. arbuscula*, 94; forest mugwort (*A. aborescens*), 81; Jewish wormwood (*A. judaica*), 81–82; *A. keiskeana*, 92; *A. lanceolata*, 94; leptofillos (*A. camprestris*), 83–84; *A. ludociana*, 94; marine (or Gallic) wormwood (*A. maritima*), 81, 97; mugwort (*A. abrotanum*), 81–82, 84, 95, 97; Roman wormwood (*A. pontica*), 81, 86, 94, 97; *A. serippium*, 97; southernwood (also mugwort: *A. abrotanum*), 81, 85–87, 97; tarragon (*see also under* tarragon; *A. dracunculus*), 81, 94; wormwood (*A. vulgaris*), 82, 84, 87, 94, 102, 143
Artemisia of Halicarnassus, 79–80
artemisinin, 93–94, 98
Asclepius (Greek:Ascelaphos), 147, see also Aesculapius
Asherah Tannit, 37
Assante, Julia, 27
Aššur, 20
Assyrians, Assyria and Assyrian medicine 16–17, 20, 26, 29, 65, 70, 82–83, 119
Atra-Hasīs Epic, 25, 34

Atropa (fate), 66
Atropa belladonna, *see* belladonna
Atropa mandragora, *see* mandrake
atropine, 58, 66–67, 142
Augustine (Saint), 39, 74, 143
Avicenna, *see* Ibn Sīnā
Avigad, Nahman, 42
axe weed, 70

Ba'al et 'Anat, 60
Baal, 20
Baaras, 64
baby snatching, 14–15, 139, 143
Babylonian Captivity, 34
Bacon, Francis, 133
Bacon, Roger, *see* Roger
Baghdad (Iraqi/ National Archaeological) Museum, 5–6
Balz-Cochois, Helgard, 22
Baquillier, Pierre, 97
barrenness, *see* fertility
Baudelaire, Charles, 109
beer, 59, 102–104
Bell, Gertrude, 6
belladonna, 66, 74
Beloch, Karl, 96
Benjamin, 56–57, 64
Bible (citations to; *see also* Genesis), 33–34, 40–41, 49, 53, 55, 57, 60, 82, 113, 131, 139, 143–145
Biggs, Robert, 25
birth, see obstetrics
Bock, Hieronymus, 89
Boeotia plate, *see* Demeter
Bogdanos, Mathew, 6
Boguet, Henry, 144
Bottéro, Jean, 19, 31
Bouchez, Colette, 75
bracken, 70
Brévart, Francis, 137–138
Brunfels, Otto, 89
Brunt, P. A., 96
Budin, Stephanie L., 27
Butenandt, Adolf, 23
Byblos, 41

cabbage, 70
Caelius Aurelianus, 71
Campari, Gaspare, 104
Camparo (drink), 104
Carpano, Antinio B., 104
Celsus, 70, 85, 95, 116–117, 122
Ceterach officinarum, see miltwaste
Chaldeans, 133
Charlesworth, M. P., 71
chaste tree, 88, 113–127, 129, 147
chichona, *see* quinine
Chinese medicine and science, 80, 92–95, 134–137, 147
Chiron, 83 (also on cover)
Chylon, 83
Circe plant, 61
circumcision and uncircumcision, 113–127
citron, 40
Claudianus, 45, 47
Clement of Alexandria, 46, 131
Cleopatria, 9
coins, 43 (Judaic), 48 (Cyrenean), 90–92 (Selinean)
Columella, 61, 70, 100–101
Commission "E" (German), 124
Conium maculatum, see hemlock
Constantine the African, 87, 96–97
contraception and abortion, 8, 10, 16–20, 23–26, 28–31, 43–44 (in Genesis), 48–51 (in ancient medicine), 84, 86; 118 (male contraceptive)
Convolvulus ovalifolius, 68
Coppens, J., 36
Corpus Hermetica, *see* Hermes, Hermetic literature
Courbet, Gustave, 109
Culpeper, Nicholas, 87, 104
cuneiform writing, 9, 26
Cyrillus, 71
Cyrus the Great, 34

date palm, 16–17, 23–24
David and House of David, 56, 76
de Ganay, Germain, 134
Degas, Edgar, 109
Della Quercia, Jacopo, 51
Demeter, see also *Hymn to Demeter*, 19, 44–47, 47 (vase of), 79, 147
demography, 145
Demosthenes, 71
depression (mental), 87
Descent into the Underworld (Inanna), 11–13
Devadasi culture, 30

Devil, 16, 65, 77, 136–137, 139, 141, 143–146
Dinah, 55, 83, 138
Diocles of Carystos, 69
Dioscorides, 24, 48–49, 61–62, 68–70, 83, 97, 102, 117, 120–122, 124, 138
Donne, John, 77, 147
Dowson, Ernest, 109
drug dealers (*pharmakopōlai*), 62
drug effectiveness, 3
drunkenness, prevention of, 86
Dubied, Danil-Henri, 104
duda'im, 60
Duke, James, 66
Dumuzi, 6, 9–10, 12, 21–22, 42
dynamis, 131, 133

Ea, 35
Ebers Papyrus, 23–24, 49, 59–60, 82
Edward the Elder, 146
Egypt and Egyptians, 37, 44, 48–49, 119, 130–135
elephant, 72–73
Elephantine, 59
elixirs, 134–137
Empedocles, 90
En, 8–9
Enki, 10, 12–13, 21, 59–60
Enkidu, 22–23
Enuma elish, 15, 34
Epic of Gilgamesh, see Gilgamesh and Gilgamesh Epic
epilepsy, 88
Epimedion, 70
Epinoia, 62
epispasm, 114–117
epistēmē, 130–131
Erech, *see* Uruk city of
Ereshkigal, 12–13, 34, 44, 47
ergotism, 24–25
Eridu, 10
estrogen and estrogen compounds, 23, 30, 50
Euripedes, 46, 80
Eusebius, 27
Evans, J. M., 39
Evans, John, 51
Eve, 6, 23, 26, 33, 36–37 (derivation of term), 39–40, 47–53, 72–73, 136–137, 141 (picture of), 147

fennel, 104
fertility (importance of, promotion, barrenness), 57–58, 68–77, 85
fevers, *see also* malaria, 95–97
Fifty Two Remedies, 93–94
flood (Sumerian; Jewish, 25, 34–35
Fowden, Garth, 132
Frontinus, 71
Fuchs, Leonhart, 89

Galen, 48, 69–70, 85, 120–122; pseudo-Galen, 117, 137
Galenic drug theory, 137
Garden of Eden, 6, 33–40, 45, 72–73, 136–137
garlic, 69
Gauguin, Paul, 109
Ge Hong, 92–94
Genesis, 6, 16, 33–37, 55–58, 60, 72–73
Geoponica, 99–100
Gerard, 123–124
Gerard of Cremona, 45
giant carrot, 116
Gilgamesh and Gilgamesh Epic, 11, 13–14, 16, 22, 34, 37–38, 51
gin, 102
gnosis, 130–131
Gnotics, 130–131
God of Wisdom, 12
Gratian, 145
Graves, Robert, 71
Grieve's Herbal, 58–59
Gula, 29

Hades, 44–45
Hadrian, 122
Hagar, 57
Hallo, W. W., 22
hallucinogens, 80, 103–111
harlot, *see* prostitution
Harry Potter, 56, 60, 68
Hathor, 59, 62
Hatshepsut, 49
Hebrew prophets, 27
Heinsohn, Gunnar, 144
Hemingway, Ernest, 108
hemlock, 66, 142
henbane, 66
Heraclitus, 134
Hermes, 44–45, 73, 120, 125, 127, 129–134, 147

Hermetica (Hermetic literature), 129–134
Herodotus, 26–27, 31, 99
Hesiod, 126
Hesychius, 60, 71
Hildegard of Bingen, 64–65, 73, 88
Hincmar of Rheims, 145
Hippocrates and Hippocratic writings, 3, 48–49, 67–68, 85, 90, 95–96, 124–125
Hobbs, Christopher, 123
Holtzman, Robert, 67
Homer, 61, 123, 126
Horace, 99, 130
Huang Di (Yellow Emperor), 92
Hugo, Victor, 112
huluppu tree, 5–31, 33
humoral balance, 137
Hymn to Demeter, 44, 46
hyoscyamine, 66
Hyoscymus niger, *see* henbane
Hyphas, 91

Ibn Sīnā, 45, 86, 96
Illithyla (or Eilethyia), 79
Inanna, 6–17, 21–22, 26, 30–31, 33, 36, 49–51, 65, 73, 129, 147
Inanna and the God of Wisdom, 12
Inanna and the Huluppu, 10–17
India and Indians, *see also* Vedic medicine, 30
insecticide, 87, 98–101
intestinal worms, *see* parasites
Iraqi Museum, *see* Baghdad
Isaac, 76
Ishmael, 57
Isidore, 71
Issachar, 55
Ištar, 19–20, 31
ivy, 70

J or Jahwist component of Genesis, 34–38, 50, 53
Jacob, 57, 76
Jacobi, H., 23
Jean Provost' guide, 121
Jesuit powder, *see* quinine
Jesus, 76
Jewish wormwood, *see* artemisia
jimson weed, 142
Jochin and Bo'az, 41
Johansen, Prytz, 46
John XXII (Pope), 145
Joseph, 56–57, 64
Josephus, 40, 64, 115
Judah, 55
Julius Caesar, 71
juniper, 69
Juvenal, 48

Kahun Papyrus, 49, 60
Kakis, Frederick, 75
Kilmer, A. D., 25
King, Leonard, 29
Kottek, Samuel S., 74
Kramer, Samuel Noah, 10–11, 24

Lachesis (fate), 66
Lamia, 139
Leah, 55–57, 62, 64, 74, 76
Lemaire, A., 42
lemon, 40
Leto, 80
Levi, 55
Levsin, 66
Life of Adam, 40
Lilith, 14–15, 26, 129, 139, 140–141 (pictures of), 143
Liu An, 92
Lotis, 120
lotus plant, 38
love apples, 19–20, 60, 65
love potions, *see* aphrodisiacs
love temples, 25–26
Lucian, 71

Maccabean rebellion, 117
Maccabee, Maccabeans, 43, 115, 117, 121–122
Macer, 86, 96, 97, 139 (German trans. of)
Machiavelli, 77
Macrobius, 71
Madaus, Gerhard, 124
magic, 135, 146, *see also* witches and witchcraft
Magister Salernus, 87
Maharbal, 71
Maia, 125
Maimonides, 86
malaria, 67–68, 80–81, 85, 89–102
Mami, 36

Mandragora officiarum; M. vernalis, see mandrake
mandrake, 55–77: 55–60 (Biblical), 59 (Egypt), 59 (beer), 61–64 (classical and medieval), 65–71 (ancient medicinal actions), 73–74 (as poison), 73–77 (fertility enhancer), 66–67 (modern science), 72–73(as fruit of Tree of Knowledge), 129, 142, 146–147
Manet, Edouard, 109
Manniche, Lise, 59
Marbode of Rennes, 86
Marcellus Empiricus, 48
marriage, 8–10, 27–28
Maspero, Gaston, 59
Mausolus, 79–80, 138
mayapple, 58
McKenzie, John L., 37
medicus, 134
melancholia, 87
melissa, 104
menstrual blood, 70
menstrual problems, *see also* contraception and abortion; premenstrual syndrome, 84–86
Michael Glykas, 72
Midrash, 55
mikku, 14, 23
miltwaste, 70
Mina (Saint), 76
mint, 70
Mishnah, 24, 50, 52
Mitchner, James, 113–114
moly plant, 61
monk's pepper, 121
Moreau, Gustave, 109
mosquitoes (incl. nets), 89
mulehoof, 70
Murray, Margaret, 145
Muthmann, Friedrich, 20
Mylitta, 26

naditu, 29–30
Namtar, 64–65
narcotic, 67, 80
Nasturium, *see* nosesmart
Nebuchadezzar, 34
Nergal and Ereshkigal myth, 25, 34
Neti, 11
nightshade, 142
Nin-gis-zida, 41
Ninhursag, 21
Ninurta, 14–15
Norris, Pamela, 52
nosesmart, 69
nu-gig, 30

oak, 16–17
obstetrics, 29, 36, 80, 85
On the Fifteen Stars, 132–133
opium, 19
orange, 40
Oribasius, 48, 121, 123
Oridinaire, Pierre, 104
Ovid, 15, 47
owl, 15, 30, 139–140 (picture), 146–147

Paestum, 101–102
Pagels, Elaine, 39
palm, *see* date palm
Pan, 120
Paracelsus, 9, 104, 137
parasites (intestinal), 82–86
Paul (Saint), 16, 52–53, 115
Paul of Aegina, 121, 123
Paulus Seilentiarius, 99
Pausanius, 126
Pepper, *see also* monk's pepper, 70
Pernod, 104
Persephone, 15, 36–37, 44–47, 51
Pharmakopōlai, *see* drug dealers
Philips, John, 52
Philo, 40, 71
philosopher's stone, 135
physicus, 134
Physiologus, 72–73
Picasso, Pablo, 109
Pireene-Delforge, Vinciane, 27
placeboes, 19
Plasmodium malaria, 67, 89–90; *P. falciparum*, 89–90, 92–94; *P. vivax*, 93
Platearius, 87
Plato, 71, 129, 134
Pliny the Elder, 48, 70, 88, 122, 123, 131
Plutarch, 71
Pluto, *see* Hades
Podophyllum pellatum, *see* American mandrake
poisons, 71–72
Polyaenus, 71

pomegranate, 5–10, 16–20, 23–24, 27, 30, 33–45 (as fruit of tree of knowledge), 44–47 (Persephone story), 48–50 (medical uses), 51–52 (in art), 65 (aphrodisiac), 129, 147
Ponchone, Paoul, 111
Pontine Marshes, 101
poplar (white), 70
Powell, Marvin A., 16
premenstrual syndrome (PMS), 84, 124
Preuss, Julius, 74
Priapus and priapism, 119–120
Procopius, 31
Propertius, 99
prostitution, 25–31, 34
pseudepipgrapha, 131
Pseudo-Apuleius, 62–64, 83
Pteris aquiline, *see* bracken
pukku, 14, 23
Punica granatum L., *see* pomegranate

qinghao, 93
quartan fever, *see also* malaria, 67–68, 89–90
quince, 40
quinine (cinchona), 96, 102
Quintus Serenus, 48
Qur'an, 36, 102

Ra, 59, 130
rabbit rennet, 70
Rachel, 55–57, 62, 64, 74, 76
Rây, Acharya, 135
Reuben, 55, 60, 62
rhizotomoi, *see* root gathers
Rimbraud, Arthur, 109
Rimmon vase, 42
rimôn, 40
rocket (*Eruca sativa*), 117–118
Roger Bacon, 129, 136, 141
Roman wormwood, *see* artemisia
root gathers (*rhizotomoi*), 62
Rowling, J. K., 56
Rubin, Jody, 116
rue, 69, 100, 117–118
Runeberg, Arne, 145

sacred marriage, 21–22, 46
saffron, 31
Salt (goddess), 103
Sarah, 57–58
Sargon, I, 31; II, 20
Šaršar, 15
Sarton, George, 135
Scarborough, John, 67
Schmandt-Besserat, Denise, 9
schools, 29–30
Schultheiss, Dirk, 116
scobas, 15
scopolamine, 66
Scribonius Largus, 48
Securigira securidaca, *see* axe weed
sedative-narcotic action, *see also* soporific, 65–67
Selinus, 79, 90–92, 98
sĕphînâytâ, 74
serpent, 13–14, 16, 23, 33, 37, 41, 136, 139
Service, Robert, 109, 147
Shakespeare, 56
Shalmenser III, 20
Shen Gua, 93
Siduri, 38–39
silphium, 48
Simeon, 55
Sin, 25
Sin-Annar, 20
Skinner, John, 39, 56
snake, *see* serpent
Solomon pomegranate vase, 42–43
Song of Songs (or *Solomon*), 21, 40, 60
Sons of El, 34
soporific, 67–71
Soranus, 48, 69–70, 85, 116, 125
southernwood, *see* artemisia
Speiser, E. I., 36
spermatogenesis, 118
spleen, 85, 90, 96–97, 104
Spot, Jean, 144
St. Anthony's Fire, *see* ergotism
St. John's girdle, 88
Stannard, Jerry, 61
Steiger, Otto, 144
Stove (goddess), 135
Strabo, 27
Strobaeus, 132
Süan, 135
Sumerians and Sumer, 9–17, 27–28, 37, 65
Sweet Annie, see *Artemisia annua*

Syriac Book of Medicines, 73–74
Syriac Herbal, 96

Tabula, 87
Taddeo Alderotti, 136–137
Talmud, 23–24, 36, 50, 52
Tammuz, 42
tarragon, 81
Tat, 133
Taweret, 59
Temple (Jerusalem), 41–43
tertian fever, *see also* malaria, 89–90
Testament of Moses, 116
Teti, 113–114
Thapsia garganica, see giant carrot
thelygonon, 70
Theobaldus, 72
Theodoretus, 71
Theodorus Priscianus, 48
Thesmophoria, 27, 45–47, 122–121, 123
Thompson, R. Campbell, 24, 64
Thothmes I, 49
thujone, 99, 106, 108–109
Toulouse-Lautrec, Henri de, 109
Tree of Knowledge (or, Good and Evil), 33, 35, 39–40, 43, 72–73
Tree of Life, 38–39, 41
tribes of Israel, 55–59
Tusser, Thomas, 100
Tutankhamun, 49, 60

Ur, 22–23
Ur-Nammu, 23
Uruk city of, 2, 5–6, 13, 31; period of, 5–8, 10, 17, 33; vase of 5–9, 16, 21, 26, 46–47
Ut, 14
Utanaphishtim, 34

Van Arsdall, Anne, 62
Van Gogh, Vincent, 109–110, 147
Varro, 100
Vedic medicine, 80, 86, 120
Verlaine, Paul, 109
vermouth, 104
Viagra, 113, 127
Vita Adae, see *Life of Adam*
Vitex (*Vitex agnus castus* or chaste tree), 88, 113–127; *V. negundo*, 118–119, 119
von Staden, Heinrich, 123

Walafrid Strabo, 88, 96, 103
Wallace, Howard N., 39
Warka, *see* Uruk
wedding customs, 8–10, 26–27
Weintraub, Maxim, 142
Wermuda, 87, 102
Wermut, 104
Weyer, John, 142
White, Lynn, 35
white wormwood, *see* artemisia
Whitlock, Brian, 75
Wilborn, Robyn, 75
Wilde, Oscar, 109
wine, 69–70, 85–87, 96, 103–111
witches and witchcraft, *see also* Lilith, 129, 139–147
Withania somnifera, 61
Wolkstein, Diane, 11, 13
wonder drugs, 137–139
Wonnecke von Kaub, Johann, 138–139

Xenophone, 70, 82

Yehonatah, 43
Yellow Emperor, *see* Huang Di

Zebulon, 55
Zeus, 44–45, 71, 79, 120, 125–126
Ziusudra, 34
Zola, Emie, 109